THE SCIENCE OF
TIME
TRAVEL

THE SECRETS BEHIND
TIME MACHINES, TIME LOOPS,
ALTERNATE REALITIES,
AND MORE!

ELIZABETH HOWELL, PhD

Skyhorse Publishing

Skyhorse Publishing books may be purchased in bulk at special discounts for sales promotion, corporate gifts, fund-raising, or educational purposes. Special editions can also be created to specifications. For details, contact the Special Sales Department, Skyhorse Publishing, 307 West 36th Street, 11th Floor, New York, NY 10018 or info@skyhorsepublishing.com.

Skyhorse® and Skyhorse Publishing® are registered trademarks of Skyhorse Publishing, Inc.®, a Delaware corporation.

Visit our website at www.skyhorsepublishing.com.

10 9 8 7 6 5 4 3 2 1

Library of Congress Cataloging-in-Publication Data is available on file.

Cover design by Daniel Brount
Cover illustration by gettyimages

Print ISBN: 978-1-5107-4964-1
Ebook ISBN: 978-1-5107-4965-8

Printed in the United States of America

To my husband, J, whose love transcends time and space;
Also, to my students, who show me what the future will really look like.

CONTENTS

INTRODUCTION

What would happen if we could travel in time?

Most of us have probably thought about this in our lives. Who wouldn't want to go back to the past and erase that regret or horrible action, like the protagonist does in the Stephen King novel *11/22/63*? Who wouldn't want to travel to the future, to see how our family turns out like in *Back to the Future*, or humanity in general like in *The Time Machine*? This book explores time travel through science fiction, because storytelling is the way that we have explored everything through the ages—from philosophy to history to creating a better world. We transmit information in the hopes that things will improve for our descendants and our society. It is a privilege to write this book, to know that these words I am writing in 2019 will be transmitted to you in some way in the future, so thank you in advance for taking the time to browse.

You can read this book in any order, although you might have better luck starting at the front. There we will be exploring the nascent stories of time travel, which dealt more in dreams than in actual machines. But as the Industrial Age shook our society, we did indeed use literal time machines in stories. All of this is covered in Part I: Old-School Time Travel (page 1), which roughly covers the late eighteenth century to the early twentieth century. Next, there are the franchises and the direct ancestors of today's stories. I'm talking the classic time-travel paradoxes of *Star Trek*, and the universe-crossing shenanigans of *Doctor Who*. I'm talking the butterfly effect and Isaac Asimov. These stories of the 1950s and 1960s take place during an explosion of science fiction interest, although at times there isn't a lot of science thrown in—at least yet. You can check this all out in Part II: The Golden Age (page 43).

Part III: The Blockbuster Era (page 91), covering roughly the 1980s and the 1990s, explores some of the time-travel stories that you are familiar with, and that indeed still are appearing in theaters today. *Terminator* was an instant classic in 1984, and has spawned so many sequels that it's been a joy to see the universe spread out. *Bill & Ted's Excellent Adventure* is expected to generate a second sequel in the near future, giving more adventures of our phone booth–traveling heroes. And here we will also talk about some newer classics of time travel, like the works of Douglas Adams and the ever-quoted *Groundhog Day*. And then there's Part IV: Into the Modern Era (page 145), which nabs some of the stories of the 2000s and beyond. One small book can't cover all the time travel of the past twenty years, let alone the past three centuries, so we know there are some gaps in this list, but here we will show you a few stories of note, including the clever *Looper* that shows how to deal with paradoxes in time, the dark *Deja Vu* that previews our struggles with cybersecurity, and the X-Men franchise, which tackled time travel in *Days of Future Past*.

For each story, I've presented a brief summary of the time travel used and (as scientifically as possible) discussed what was going on. I also delved into a related theme that I think might be interesting to younger readers, especially as they try to find their way in life. After all, what's a good story without trying to apply it to your workplace or family or to improve the world? I hope to give you some points to think about as you ponder these questions. And finally, I delve into the legacies of each work. Sometimes they appear small, sometimes they loom large, but overall, they each have a unique contribution in the time travel literature.

If you really enjoy the stories in this book, I invite you to learn more about the authors and the time periods in which they wrote. The summer I wrote this book (2019), I was lucky enough to have a paid-for business trip in the city of Dallas, the famous scene of the Kennedy assassination in 1963. Parts of the block are preserved just as they were on the day that he died. And people all over the world still flock to see it, overcrowding the famous Sixth Floor Museum that chronicles his life and times. On a weekend morning, I wandered the area and it sure did bring to mind the book *11/22/63*. But it also made me think about time travel in general. Some events seem timeless, even though they took place long ago. The thousands of people I saw honoring Kennedy's legacy reminded me that time-travel stories do have resonance. And to paraphrase T. S. Eliot, it's only by moving through time—returning to the place where we began—that we can truly understand the journey.

Thanks for sharing your time with me!

PART I
OLD-SCHOOL TIME TRAVEL

ANNO 7603
(1781)

This kicks off a series of older tales about time travel, for which I will warn: the scientific explanation for most of the time travel in this section is not really "scientific" in the sense that we understand it today. That said, *Anno 7603* is one of a series of stories I chose because their plots have significance to understand not only today's time-travel stories, but also today's culture.

The Science of Time Travel

The premise of this Johan Herman Wessel tale is a fairly simple one, although we have to suspend disbelief in our modern sensibilities and try to read the story with an understanding that things were different a couple of hundred years ago. Julie and Leander are in love with each other and are interested in how it would be to exist as the other's gender, specifically with the eye of raising children better, so a fairy magically takes them forward in time to the year 7603 to see what would happen. Therefore . . . yeah, there isn't much science here to be had in terms of discussing the time travel trope. Rather, let's focus on what this visit reveals.

Without focusing on many of the story's spoilers, one thing that they discover is only women are allowed to fight in the military. Let's remember that this story was written long before a time when women were allowed to vote or to run their own households, let alone make decisions that could put themselves in danger. Even today, women face discrimination in military spheres such as jet piloting or submarining, where they are in a very small minority. The romantic pair conclude their journey through time traveling not only deciding that it's best not to switch genders, but also that they are more prepared to act in their societally mandated roles after looking at the alternatives. It doesn't sound like a traditional hero of a story, in the sense that they're not prepared to muster change. So while this story is an interesting look at how the role of gender could influence society, it feels somewhat incomplete in reading it.

Many mythological stories in the centuries and millennia before this one follow a somewhat scripted formula that was later identified by Joseph Campbell as "the hero's journey." The main thing to understand here is that the hero is supposed to return from their journey somehow changed. One can argue that this couple did return changed—after all, they no longer want to switch gender roles—but their change didn't have broader implications for society.

If we think about modern examples of the hero's journey, we can definitely see character change happening as a result of it. George Lucas (of *Star Wars*) was greatly influenced by Campbell's work. If you look at the character of Luke Skywalker in the original trilogy, he starts the series as a somewhat-whiny nephew of his aunt and uncle, complaining about not being able to participate in the war and of having to follow his elders' orders. Along the way, he meets characters who have been to war and who have lived through some of its terrible implications—people dying, people being captured, people losing their homes forever. By the original trilogy's end, Skywalker does become a novice warrior, but not a warrior seeking adventure—rather, a warrior seeking to help others. (We'll let you watch the sequels to find out what happens next.)

The Science of the Scientific Method

I mentioned earlier that the concept of fairies taking people on a journey is not scientific, and I stand by that. That said, fairies are an example of how people used to explain the unexplainable. The scientific method as we understand it is just a few centuries old.

So, what is the scientific method? It's actually something that you can use in your own life, every day. Do I work on my studies better in the morning or the evening? Is it faster for me to get to school or college by bus, or to bike? A good way to solve these problems is through experimentation. You repeat the process a few times, try changing around the variables, and then observe the results. You may then have to revise your hypotheses (your initial ideas) depending on what the conclusions are. But the point is, you never assume and you always test it.

While the name "scientific method" dates from about the nineteenth century, we can see its roots in the scientists and philosophers of a few centuries beforehand, including Isaac Newton (who helped derive our modern understanding of gravity), Francis Bacon (an early figure of the scientific revolution and "empirical" or evidence-based thinking) and Rene Descartes (who used reason to examine the natural sciences).

A simple way of thinking about the scientific method comes from ScienceBuddies[1]:

1. Ask a question
2. Do background research
3. Construct a hypothesis
4. Test with an experiment, making sure that the procedure works, and adjusting as needed
5. Analyze data and draw conclusions; the results either align with the hypothesis, or they do not
6. Communicate results

Let's take a brief look at each of these points, and why they are important:

---

1 ScienceBuddies. (2019). "Steps of the Scientific Method." Retrieved from https://www.sciencebuddies.org/science-fair-projects/science-fair/steps-of-the-scientific-method

1. Ask a question: You can't run an experiment unless you know what you are testing for, so you need to find a question that is testable with the materials you have at hand.

2. Do background research: This is to find out who else has been interested in your question, or related questions. Make sure to cite these people carefully as you go through the experiment.

3. Construct a hypothesis: Take a position on what the results will be, and why they are important in the realm of science. You will best be able to construct a hypothesis while looking at the literature.

4. Test with an experiment, making sure that the procedure works, and adjusting as needed: This sort of skips over how to make a good experiment, but it does emphasize that the way that we do it is important and that we should try different variables to ensure we have the right design.

5. Analyze data and draw conclusions; the results either align with the hypothesis, or they do not: Here you not only analyze the result, but also try to figure out why you got it, and the implications for future research.

6. Communicate results: Science shouldn't take place in an intellectual vacuum. Share your experiments and ideas in some way (through a paper, a website, a social media feed) and see how the community responds. Then they can build upon the knowledge for the next thing.

While the high school environment (or even the university environment) makes these steps feel like busywork, they are very important in figuring out all sorts of problems in everyday life. One of my favorite examples in literature of the scientific method in use is a passage in *Zen and the Art of Motorcycle Maintenance* where the narrator uses it to explain how to solve sticky motorcycle issues, such as a stuck screw. But if you look carefully, you'll find the scientific method's use everywhere.

The Legacy of Wessel

Wessel's legacy with this particular story, admittedly, is not great. In translation, the premise seems confused and the ideas that he has about society and gender are outdated. The story is not often cited among his works, presumably because it is difficult to understand and the characters don't undergo much growth during the story (making it perhaps uninteresting to the reader, although it's hard in modern times to say what somebody in the eighteenth century would have enjoyed). However, it's notable to realize that in his short forty-three years, he made an impact on those who were defining Norwegian identity in the time, especially through the use of literature. He was a prominent part of the Norwegian Society, who worked in Copenhagen to figure out how to portray Norway in poetry and other literary works.

Modern scholars of Wessel will appreciate that he wasn't afraid to make fun of society—he was humorous and satirical. One of his most famous tales was Smeden og Bageren ("The Smith and the Baker"). In this story, a case of manslaughter is being considered and it is decided to execute the baker instead of the smith, even though the baker has the stronger alibi. Why? It is said that the smith has the more "important" role in society. Today, that comparison still rings true. How many of us have seen cutbacks in the arts or sciences in the name of making a bit more money?

To move back to the Norwegian Society, this was a group of (mostly) men—a common feature of discussion groups at the time—that had an eclectic mix of philosophers, authors, and poets. Founded in 1772 by government official Ove Gjerløw Meyer, the group wanted to bring more of an awareness of Norwegian patriotism into society. One of their major contributions was influencing the creation of the first Norwegian university. As closed-door as this group is, it is a great idea when people can get together informally or formally and try to figure out how to improve their society to make a greater impact for the good. It's something that many of us can do quite easily and at low cost, through community improvement groups, libraries, and schools, to name a few. Mind you, not all of us can afford the time or the energy to take part in these conversations and to effect change. But if you can spare a few hours of volunteerism every year or every month, you'd be surprised at just how far a little effort can take you with the right engaged group working with you.

RIP VAN WINKLE
(1819)

Man goes into the mountains, man falls asleep, man emerges . . . in the future? This somewhat confusing tale is something that most of us already are aware of in our cultural upbringing (if we come from the United States, in certain groups), but it has a wider discussion embedded about the American Revolution and its meaning for science.

The Science of Time Travel

You've probably heard of the premise of Rip Van Winkle, but it turns out this simple children's tale has a fair amount of history buried beneath it. Essentially, Van Winkle is living in colonial America as a Dutch-American villager. He goes into the mountains one day (in part because he feels henpecked by his wife, which is a gender issue that we'll note but not dwell upon), falls asleep, and wakes up. What few retellings of the story mention is that he happens to wake up after the American Revolution took place, so obviously there's a great societal shift that Van Winkle finds himself struggling to cope with.

One of the most obvious manifestations of the change comes when Van Winkle starts telling his villagers (none of whom recognize him) that he is loyal to King George III. That was (to say the least) an unpopular position given that at one point, the leaders of the revolution were declared traitors in Britain, and that George III is sometimes accused of continuing the war even against his ministers' recommendations. In any case, our friend Van Winkle is finally recognized by somebody else in the village and his potential for getting into trouble quickly diminishes.

How does Van Winkle travel through time? Well, unfortunately, again we don't have a lot of science to help us. Essentially he drinks something from a keg offered by strangers, which reminds him of his homeland since it had "the flavor of excellent Hollands." As Washington Irving writes, "He was naturally a thirsty soul, and was soon tempted to repeat the draught. One taste provoked another; and he reiterated his visits to the flagon so often that at length his senses were overpowered, his eyes swam in his head, his head gradually declined, and he fell into a deep sleep."[1] So, what exactly happened on the mountain? Did the keg put him to sleep for twenty years? Again, that's not very well explained. A local historian is consulted near the end of the story and tells the audience "that the Kaatskill mountains had always been haunted by strange beings." It turns out that the strangers Van Winkle encountered were the ghosts of the explorer Henry (or Hendrick) Hudson and his crew, who had died at sea. In the tale, they show up in the mountains as ghosts every twenty years, "being permitted in this way to revisit the scenes of his enterprise, and keep a guardian eye upon the river, and the great city called by his name."

If we're to try to make any sense of this at all, we can talk about the idea of different people experiencing different frames of time. We'll talk later in the book about the works of Albert Einstein, the great scientist and mathematician of the twentieth century who revolutionized how we think about the universe and space-time. His works suggest that traveling near the speed of light, people can experience different time frames. Somebody going at that ultra-fast speed would experience time moving more slowly, specifically, than somebody living on Earth. So, it would be possible for a person to go on a journey far into space, travel for a few years, and then return home discovering that the "few years" was actually many

---

1   Irving, Washington. (n.d.) Rip Van Winkle, a Posthumous Writing of Diedrich Knickerbocker. Bartleby.com. Retrieved from https://www.bartleby.com/310/2/1.html

centuries or even more on Earth. To paraphrase the 1997 movie *Contact* (which touches exactly upon this theme when somebody suggests going to visit aliens in a distant galaxy), everyone that you know will have been dead and buried.

Clearly, this story was written before the birth of Einstein, and also as clearly, Van Winkle was not traveling at the speed of light. In fact, he wasn't traveling at all, but lying asleep in some sheltered area and miraculously remaining undiscovered by the villagers for decades. This idea of somebody going to sleep and waking up again many years later is a common theme of literature, even in the fairy tale *Sleeping Beauty*. What makes this one stand apart, however, is that Van Winkle wakes up and has traveled through time, at least in the sense that he didn't experience the last twenty years.

The Science of the American Revolution

The American Revolution is so ingrained in history lessons in the United States that it's sometimes hard to separate the mythology from the reality. As we all know, its origins came in large part due to dissatisfaction with British rule of the thirteen American colonies, including the famous "no taxation without representation" cry that came because Americans had no position in the British Parliament.

The roots of the revolution not only come from historical events, but also from changing in thinking in the colonies. Specifically, there were a number of people who participated in a field of philosophy known as the American Enlightenment (which, ironically, was influenced by the Enlightenment movement in Britain). This was the idea of applying the scientific method and scientific reasoning outside of the ideas of science—particularly, to politics and religion. This was less a time of believing in a trusted authority, and more a time of trying to follow logical threads and coming to a conclusion. (For more on the scientific method, consult the chapter on *Anno 7603* on page 3).

As an example, let's talk about the brilliant Sir Isaac Newton, who created the laws of motion and the law of universal gravitation, which help us understand how the solar system's planets orbit the sun and how to send spacecraft between planets. "Seeing this amazing success, the philosophers of Newton's day asked: If there are natural laws for how planets move, shouldn't there be natural laws for how people should act as well?" writes the *Independent Record*.[2] "And if so, what are these natural laws? Perhaps the most important philosopher to consider this question was the Englishman John Locke. Locke carefully studied Isaac Newton's new theories, and the two men became friends, meeting and writing to each other about a variety of issues." As the publication explains, each of these people had ideas that influenced the American Revolution. Locke, for example, was greatly influenced by natural laws, arguing that because of them, every person should have the ability to determine their own destiny.

The Legacy of Rip Van Winkle

One way in which Rip Van Winkle remains culturally relevant is in the numerous adaptations of the story. We've seen it in theater, music, film, television, cartoons, comics—you name it. Perhaps you can think of it as an early science-fiction franchise, in the spirit of today's Avengers or X-Men. And lest you think that it is a dated franchise, it showed up (in all things) in the 2018 video game Red Dead Redemption 2, when one character makes a sarcastic remark identifying himself as Rip Van Winkle.

When Irving's birth passed its 200th anniversary in 1983, the *New York Times* pointed to the amount of influence he had on the Hudson Valley in his lifetime. "Washington Irving, certainly the

---

2   Cline, Kelly. (2011, May 4). "How Isaac Newton's science inspired the American Revolution." *The Independent Record*. Retrieved from https://helenair.com/lifestyles/health-med-fit/how-isaac-newton-s-science-inspired-the-american-revolution/article_ea6a77b0-760c-11e0-bfc4-001cc4c002e0.html

county's most famous resident, became one of the nation's most beloved and respected citizens and is considered by many as this country's first genuinely American author. The village of Irvington renamed itself in his honor during his lifetime and by the time he died in 1859, both the man and his house in Tarrytown were almost national shrines."[3] It appears that nearly forty years later, at least some residents of the Hudson Valley still feel some affection for Irving despite him having passed away ages ago. But there is an acknowledgment that he might be losing resonance with the passing of time.

"I read more and more of Irving's works and then on to the biographies of Irving. It became clear that this writer, once the most popular of Americans, who helped to forge a young democracy—its laws, its values, its sense of history, its art and culture—had been forgotten. How does that happen?" asked a writer in *The Hudson Independent*.[4] While the writer didn't offer any clear answers (instead citing cultural references from New York's center of gravity status, to the American Romantic movement that Irving influenced), perhaps we can draw some wisdom in knowing that good stories are timeless, but their challenge is finding an audience. And sometimes, the audience simply changes enough over the ages that the cultural influence is not the same as it was before.

I'd like to think that one of the marks of progress of a society is throwing off old-fashioned ideas that are put forward to the detriment of other groups. In recent years, most of us older ones have shed our past misguided thinking about matters such as homosexuality or women's rights, among other things. And while Irving still has cultural influence in the matter of somebody being

---

3   Melvin, Tessa. (1983, April 3.) The Legacy of Washington Irving. *New York Times.* Retrieved from https://www.nytimes.com/1983/04/03/nyregion/the-legacy-of-washington-irving.html

4   Moffat, Lynn. (n.d.) What The Legend Means to Me. *The Hudson Independent.* Retrieved from https://thehudsonindependent.com/what-the-legend-means-to-me/

behind the times, parts of his story are cringeworthy; not only does Rip Van Winkle not enjoy being around his wife, he actually expresses relief when he finds out (upon his return to the village) that she has died, implying that he felt that his freedom depended in large part on not having her around. Ideas change, times move on, but it's also important to know where we came from. And for that reason, I do hope that even people outside the Hudson Valley appreciate the importance of Rip Van Winkle.

THE CLOCK THAT WENT BACKWARD (1881)

inally, we move to a story where a machine is used for time travel. This isn't a voyage to the future in the modern sense, but this child's story shows us one of the earliest uses of mechanical time travel. This shift to machines (over dreams) shows the mark of industrialization starting to make its way on societies around the world.

The Science of Time Travel

This short story by Edward Page Mitchell focuses on the "periodical visits of duty" that two boys, the narrator and his cousin Henry, have while visiting their Aunt Gertrude in Maine. Like in *Rip Van Winkle*, there's a lot of discussion in the story about Dutch ancestors and the early colonial days of the United States, although the narrator is clever enough to portray this family history as boring, due to "the constant repetition and the merciless persistency with which the above dates were driven into our young ears that made us skeptics." The boys also acknowledge that they feel all of these ancestors are "pure myths," only embodied in the stories of their aging aunt.[1]

The story then turns to a Dutch clock sitting on the first landing in the mansion, made in the year 1572 by a local craftsperson. But the clock, as can be guessed by its age, appears to be dead. Aunt Gertrude explains it away with a lightning strike (which, as you'll

---

1   Mitchell, Edward Page. (1881). The Clock That Went Backward. Retrieved from http://www.forgottenfutures.com/game/ff9/tachypmp.htm#clock

find out, sounds a little like the premise of *Back to the Future*, which we'll explore in Part III on page 109). One night, however, the boys (although they say they had "grown out of boyhood" by this time) hear some noises and leave their beds to investigate. They see their aunt beside the clock, "take a key from behind the face and proceed to wind up the weights." To their astonishment, the hands of the clock move backward in front of their eyes. Moments later, their aunt collapses in front of the clock; it's implied in the story that she is struck dead.

We'll skip forward in the story, ahead a few years to when the boys, now clearly adults, begin to piece together the mystery of what happened. In talking with a learned professor, they begin discussing the nature of time as put forward by the philosopher Georg Wilhelm Friedrich Hegel. As the story explains: "Time is a condition, not an essential. Viewed from the Absolute, the sequence by which future follows present and present follows past is purely arbitrary. Yesterday, today, tomorrow; there is no reason in the nature of things why the order should not be tomorrow, today, yesterday." Eventually, the narrator and his cousin find themselves using the clock to go back in time to the Siege of Leiden, an event of the Eighty Years' War (1568–1648) having to do with the Spanish occupation of Leiden. Without giving too much away in the story, it turns out that these visitors to the past have had an influence on the future. In fact, the professor muses upon cause and effect at the end of the story:

> If cause produces effect, does effect never induce cause? Does the law of heredity, unlike all other laws of this universe of mind and matter, operate in one direction only? Does the descendant owe everything to the ancestor, and the ancestor nothing to the descendant? Does destiny, which may seize upon our existence, and for its own purposes bear us far into the future, never carry us back into the past?

As we'll see in later stories of the book, all this discussion about the interweavings of the past, and how it influences the future, will lead to troubles such as time travel paradoxes (such as the classic about going back in time and accidentally killing one's parents or grandparents). It's also true that a lot of time travel stories attempt to make conclusions about the past from our more "modern" sensibilities, although I admire this particular tale for restraining itself in that respect. In fact, the narrator finds himself upon a war field running around in a state of confusion and getting injured in the process. This sounds like a much more realistic feeling for a new visitor to the past than being able to affect the future through a brief encounter, although other stories don't shy away from this concept.

Unfortunately, it's never quite explained how the clock puts us back in time, except to say that it is old, it is from a certain heritage, and that it is mechanical. Interestingly, the clock does seem to contain some sort of reference to gunpowder (a novel technology of the day in Europe): "At the same instant a ball of fire, leaving a wake of sulphurous vapor and filling the room with dazzling light, passed over our heads and smote the clock." But I suppose one could say that the clock opened up a wormhole, or some other such excuse for time travel that we can explore in later stories.

The Science of the Industrial Revolution

In a brief summary, it's hard to capture the importance of the Industrial Revolution, but there are a couple of aspects that we can touch on here. This period in Europe and North America ran roughly from the mid-1700s and the mid-1800s, and is most classically portrayed as a period where people moved away from using hand production to machines, which led to standardizations in machining that influenced the industry for decades to come.

While most of us remember the mighty machines of the era like the railroad and the telegraph, a lot of what happened occurred in more humble professions such as textiles and farming. The use of machines made food and other goods more widely available and at a lower cost, which slowly began to increase the standard of living in terms of buying goods. However, workers suffered in other senses, with the introduction of dangerous equipment, long hours, and the need to move into overcrowded cities to find work. Another unfortunate side effect was the increase of colonization, where raw goods were stripped from countries in Africa and South America and sent overseas (along with slaves in some cases) to the richer countries. The effects of this era are still echoing around the world today.

One of the greatest effects of the Industrial Revolution was an increase in literacy in many populations. The printing press was further commercialized and made it possible to create books at a mass scale, cheapening the cost and making it easier for people to access information. Around the same time, schools were implemented and children were allowed to learn reading, writing, mathematics, and other basic skills in preparation for the workforce.

For the purposes of time travel, the industrial revolution reduced our use of dreams and the fantastical to explain how people travel to the past and future, and replaced it with more mechanical versions. In some tales of time travel, the fact that somebody uses a machine is "enough" and we don't find out more about how the machine works. In others, especially in later years, people try to base the machine's work in more scientific principles. In any case, many of the stories we'll explore from here on rely less on dreams and fairies and mythology to move through time. One author in TechRadar argues that time travel seemed more possible after the Industrial Revolution, simply because the pace of change appeared to be accelerating. "The industrial revolution changed all of this. For the first time in human history, the pace of technological change was visible within a human lifespan," wrote James O'Malley. "It is not a coincidence that it was only after science and technological change became a normal part of the human experience, that time travel became something we dreamed of."[2] O'Malley is right that mechanical time travel is something that was best explored during the Industrial Revolution, so you can expect to read a lot more about machines and their various ways of time traveling in other stories explored in this book.

The Legacy of Edward Page Mitchell

While this short story by Mitchell isn't often cited in the literature, it is notable that it uses a machine to go back through time—and that this happened even before H. G. Wells's *The Time Machine* (which we will explore next on page 20). Some people say that Wells was influenced by Mitchell, although perhaps the safest bet is that both people were influenced by the reach of the Industrial Revolution in general, as well as intrigued for what this might mean for the implications of time.

Mitchell's work was anticipatory of a lot of trends in science fiction, actually, including faster-than-light travel, computers and cyborgs, mutants and teleportation. I think he'd feel at home in most of the Avengers plots of today. (Funny to think how much our understanding of science fiction has advanced in the last one hundred years or so, since audiences only a century ago found these ideas novel and strange, and today they are almost expected.) Mitchell was in fact a poorly known author for much of the twentieth century, as his stories were often published without bylines and were thus poorly indexed. One of the major efforts to track down his work was a 1973 anthology of his stories, compiled by Sam Moskowitz. Who knows, there might be more stories of Mitchell waiting to be found by future scholars. In general, though, we can say that Mitchell was an interesting representation of the promise of the Industrial Revolution, and of all of the time travel and science fiction stories that we enjoy today.

---

2   O'Malley, James. (2017, Aug. 2). "A history of time travel: the how, the why and the when of turning back the clock." TechRadar. Retrieved from https://www.techradar.com/news/a-history-of-time-travel-the-how-the-why-and-the-when-of-turning-back-the-clock

THE TIME MACHINE (1895)

This story is the distant ancestor of most time travel stories today. It's not only a discussion of how technology can change our lives, but is also a commentary on how our society could evolve in future ages through Darwinian selection, a new idea at the time.

The Science of Time Travel

This H. G. Wells novel follows an adventurer who is referred to simply as the Time Traveler. We know a few things about this fellow: he's Victorian, he's English, and he has a relatively independent lifestyle allowing him to tinker around with machinery. Every week, in the style of people with leisure time in the late 1800s, he would regale his dinner guests about how time works.

Here, we follow the adventures of his journey far into the future. The traveler tests his machine and manages to leap forward five hours. Feeling that's not nearly far enough, he pushes time farther forward and watches his house (over the ages) change into a garden. Finally, he stops the clock at AD 802,701.

It's here that the narrator meets the two dominant groups of people living in London around that time: the Eloi (who are childlike) and the Morlocks (who are apelike and prefer to live underground, unless it's nighttime outside). There's a bit of dated social commentary and science that I'd prefer not to tackle here due to the ethical complexity, but suffice it to say that their states of nature are supposed to be the outcome of thousands of years of human evolution. The narrator finds himself trapped in the past and fighting to find a way back to the future, while working with these two groups of people to accomplish that goal.

The machine (unlike the grandfather clock that we saw in a previous story) is something that a person can sit in. In fact, it's recognizable enough as a time travel device that it would still fit in movies today. Wells describes it as "a glittering metallic framework, scarcely larger than a small clock, and very delicately made. There was ivory in it, and some transparent crystalline substance." Inside is a seat and two simple levers to go back or forward in time.

Do we get much of a scientific explanation about how time travel happens? Well, there is much discussion in the book about how time is a fourth dimension of space, which again was a relatively new concept at the time. This was written just before Albert Einstein developed his groundbreaking theories of time and space, but there was a lot of scholarship and discussion going on about where time fits into science and how we can measure it, so how the actual traveling through time was accomplished is best left to speculation. Our best guess is that the time machine somehow generated a wormhole, which is a tunnel between two different points in space. It allows people to go faster than the speed of light (which is the cosmic speed limit, as we'll explore later on) because of a tunnel in space. Now, how that wormhole exactly was generated is an open topic, because one would assume that a wormhole that swallows a machine would be at risk of grabbing other nearby objects. Presumably the machine must open and close the gateway to the wormhole, because otherwise, if it's stuck open anything could just fall in.

Notably, this story puts forward the belief that one can travel in time *only* and not in space. Specifically, the narrator goes back in time to the same location, and doesn't switch locations at all. This makes the plot a little more believable, although again, the methodology of traveling through time is mostly left up to the imagination. While it's hard to reconstruct a scientific discussion of how time travel occurs, there are some things to note here. The first is the use of a device to travel through time, which was a fairly new idea. Wells's work became widespread and is still a large influence on time travel stories today, because we see a lot of stories where people jump into an object of some sort to go into the future (spaceships, phone booths, police call boxes, and the like). The second is a discussion of science surrounding time. Perhaps the use of the time travel machine itself was not that scientific, but there is discussion about how time works, including the use of models. You'll see some other stories in this book that try to use scientific explanations—Michael Crichton is a famous author who constantly used science to explain not just time travel, but dinosaur cloning as well. So, you can see this H. G. Wells tale as a predecessor to more scientific explanation in stories.

The Science of Evolution

One of the things that is bothersome about Wells's work on evolution is the idea that there is always something better to strive for, and that humans should be shaped by eugenics (controlled breeding) to attain a utopian end. This brings to mind so many issues about human freedom and classism that we won't even attempt to unpack it here, so instead we will focus on the basics of the modern understanding of evolution and why it is important to science.

The first thing to understand is that the world is old and the universe we live in is even older. The current numbers (which may change as we get better at dating things) show that the universe is roughly 13.8 billion years old and our solar system (including the Earth) roughly 4.5 billion. Life first showed up roughly 3 billion years ago, although that again is subject to interpretation. We can only go by what shows up in the rock record, and it's very possible that delicate forms of life came to be and couldn't be preserved, but science proceeds on the evidence, so we focus on what we can say for sure.

So, when did humans show up on the scene? Well, here's where evolution gets a little sticky. Modern-day apes and modern-day humans share a common ancestor that likely lived about 5 million to 11 million years ago in Africa, according to HowStuffWorks[1]. Since then we've split into two distinct groups (apes and humans, that is) and humans have evolved in some different ways, including walking upright and having larger heads. Some people say that evolution cannot be tested, which actually isn't true, How Stuff Works adds. "Scientists have successfully run numerous laboratory tests that support the major tenets of evolution. And field scientists have been able to use the fossil record to answer important questions about natural selection and how organisms change over time." We can even see human-induced evolution in creatures of much shorter lifespans. It's been used in everything from breeding animals to creating hardy plants for agriculture.

Evolution typically turns controversial when using it to explain so-called differences in races or in people's abilities. First of all, these aren't fair assessments, and second of all, they are largely disproved by science. When you're confronted with something (including but not limited to evolution) that challenges your beliefs, dive into the literature. Read the studies and look for those that cite a wide field of information (often referred to as cross-longitudinal studies or literature reviews). If you have a body of information that is largely citing a few things, there is more scientific weight behind it. Scientists can say that evolution exists and that the world is warming because there's so much weight behind these ideas that the evidence is indeed overwhelming. That said, always read carefully and look at the original evidence. You'll be a better student of life that way.

---

1   Harris, William. (n.d.) "Are humans really descended from apes?" HowStuffWorks. Retrieved from https://science.howstuffworks.com/life/evolution/humans-descended-from-apes.htm

The Legacy of *The Time Machine*

The first thing to note about this book's influence is how many adaptations it has inspired over the years, including at least four films, two television shows, several radio adaptations, and even comics. Incredibly, many authors have taken it upon themselves to write sequels or works that carry on from the ideas of Wells's work. One of the more recent ones was *The Time Ships*, written by Stephen Baxter and authorized by the Wells estate because it was written in 1995, a century after the story's original publication. Wells has in fact many themes that remain popular today, covering everything from interplanetary war (*War of the Worlds*) to the troubles of meddling scientists (seen in small part in *The Time Machine*) to time travel itself. One 2002 article asked where Hollywood's science fiction would be before H. G. Wells, and the author's great-grandson (Simon Wells) had a unique response. Simon was well positioned to talk about his ancestor's influence since he directed a film of *The Time Machine*. "It's like trying to say, where would theater have been without Shakespeare?" Simon Wells told The Associated Press.[2] "There were lots of other people who were writing great stuff, but somehow certain people become the touchstone, the sort of golden standard, the reference. And H. G. certainly was the gold standard in science fiction."

While earlier I criticized Wells for not describing much of the science behind his time machine, scholars have argued that it makes his work more timeless than a great contemporary of his, Jules Verne, who described his technological innovations in great detail, which included such uninvented novelties as submarines and rocketry. While he had the concepts right, he made unavoidable technical mistakes in the details because nobody knew how these

---

2　The Associated Press. (2002, March 20). "Time travels on — thanks to the rich legacy of H.G. Wells." *Deseret News.* Retrieved from https://www.deseret.com/2002/3/20/19644316/time-travels-on-151-thanks-to-the-rich-legacy-of-h-g-wells

things worked. "The way it strikes me is that Verne was much more explicit in what he was describing, and that works against him today," John Partington, editor of H. G. Wells Society's scholarly journal *The Wellsian* said in an interview with the Associated Press. "When Wells describes his time machine, he's very vague. That means it can fit into any generation. That's one reason Wells remains fresh and Verne dates."

There were aspects of Wells's work that did not age as well, however. *The Independent* notes that Wells had an idea of a utopian world state.[3] Wells argued, in his collaboratively written *The Outline of History* (1919), that human evolution would eventually outstrip the nation-state that still makes up the political sphere of our world today. One thing that Wells can be happy about, though, if he were able to time travel from the 1850s into today, is that his idea of universal human rights now only has been adopted by the United Nations, but it's something that more and more countries are agreeing to. He wanted people to have access to education; work, a place to live, heck, a right to live. And that's something that is an even more imploring call to action today. There are so many of us, even in industrialized countries, who still struggle for basics such as housing or a post-secondary education. While Wells's ideas for a utopian society have outgrown his time, it would be wonderful if we could do more to unite behind some of the principles of giving people basic rights to life.

3   James, Simon John. (2016, Sept. 22). "HG Wells: A visionary who should be remembered for his social predictions, not just his scientific ones." *The Independent*. Retrieved from https://www.independent.co.uk/arts-entertainment/hg-wells-a-visionary-who-should-be-remembered-for-his-social-predictions-not-just-his-scientific-a7320486.html

A CONNECTICUT YANKEE IN KING ARTHUR'S COURT (1889)

What science fiction story wouldn't be complete without citing medieval times? You can think of this story as a somewhat distant ancestor to the classic *Monty Python and the Holy Grail*, since both completely lampoon the Arthurian age while also showing a large respect for it.

The Science of Time Travel

I asked my virtual assistant, Christina Goodvin, to help me seek out the scientific explanation between different forms of time travel. This was her brilliant summary of this particular book: "Man gets hit on the head with a crowbar during a fight with one of his employees and awakes under a tree to find himself in the past. Returns to present when Merlin casts an enchanted sleep on him." Clearly this is . . . not such a scientific way of returning to the past, unless we're to wave it away as a dream sequence. That didn't really work for that *Dallas* episode "Who Shot J. R.?" so I'd prefer to believe that our narrator really did have a lived experience in Arthur's Court. That said, I can't prove it.

Let's back up and talk a bit about the plotline here. Our narrator is an engineer named Hank Morgan, who needs to adapt quickly when he arrives back in Arthurian times. He, of course, comes from so far in the future that he seems very strange to the locals, and possessed with special powers. So Morgan, the adaptable sort, decides to frame himself as a magician and to (perhaps unwisely) compete against Merlin in terms of impressing the populace. His first act is showing up just before a solar eclipse in the year 528, which the populace does not know about, but Morgan, having read about it in his previous life, actually does. This is handy knowledge to have when you are about to be burned at the stake due to looking and acting strange, at least by the standards of the sixth century. Morgan threatens to obscure the sun if he is harmed, and of course his prediction comes true and he is freed.

While the story is a satire, it may have roots in true history. In 1502, Italian explorer Christopher Columbus was in Jamaica and carrying a copy of an almanac constructed by Johannes Müller von Königsberg, more popularly called Regiomontanus. This came in handy for Columbus (but unfairly exploited the peoples he encountered) when his crew fell into dispute with the Arawak Indians, according to Space. com.[1] Initially the locals were generous with food and shelter, but then Columbus's crew (which was stranded in Jamaica due to issues with the ship) repaid their generosity with frustration, robbery, and murder. Understandably, the locals did not want to engage any further with the crew, but Columbus used his knowledge of the eclipse to say his god would be angry if the Arawaks refused to provide food, and blot out the moon. When the eclipse happened, the terrified Arawaks consented. It was a very mean trick played by Columbus in an era when so many indigenous groups were being exploited, murdered, and sometimes forced into slavery by explorers.

There's quite a lot of classism in the book, especially as Morgan tries to industrialize the local population. He expresses frustration with the way that the locals talk (which, incidentally, is sometimes ripped right from the pages of Thomas Mallory's *Morte D'Arthur*). He is impatient with how meandering their stories seem and how uneager they are to rush around. Morgan has this idea that he is mobilizing the people for their own good as he introduces industrialization, but in reality, he's showing a bias of progress and a feeling of superiority. Perhaps it cannot be helped given the

---

1   Rao, Joe. (2014, Oct. 12). How a Total Lunar Eclipse Saved Christopher Columbus. Space.com. Retrieved from https://www.space.com/27412-christopher-columbus-lunar-eclipse.html.

tendencies of his time when entire cultures were schooled to think they were better than indigenous peoples, but the humor fades as we think about the implications for all the oppressed peoples over the years who have been "bettered" by the more powerful—only to their detriment. In any case, Morgan tries to be at least a little sneaky in the way that he sets up a more modern society. Perhaps his most interesting idea is to create modern factories, but to carefully select only a few of the populace to go and work in them. In general, he seems to focus on people who aren't too set in their ways, unfortunately, as he prefers having workers that are easy to be with and who don't question his motives.

The Science of Gunpowder

While gunpowder wasn't used in King Arthur's court, it was used in China in the ninth century. The original concoction was mixing sulfur, charcoal, and saltpeter (also known as potassium nitrate), although there are variations on the ingredients, according to ThoughtCo. "The ingredients tended to require remixing prior to use, so making gunpowder was very dangerous. People who made gunpowder would sometimes add water, wine, or another liquid to reduce this hazard since a single spark could result in a smoky fire," the website wrote. Gunpowder (and also rocket fuel, incidentally) both require the use of a fuel and an oxidizer to react. As the website explains, the charcoal generates carbon. The carbon's reaction with oxygen creates carbon dioxide and energy. Since carbon in a fire tends to suck out oxygen from the air, saltpeter provides an extra oxygen boost to speed up the reaction. "Potassium nitrate, sulfur, and carbon react together to form nitrogen and carbon dioxide gases and potassium sulfide," the website adds. "The expanding gases, nitrogen and carbon dioxide, provide the propelling action."[2]

---

2   Helmenstine, Anne Marie. (2019, June 6). "Gunpowder Facts and History." ThoughtCo. Retrieved from https://www.thoughtco.com/gunpowder-facts-and-history-607754

The Legacy of *A Connecticut Yankee* in *King Arthur's Court*

While some of the book's humor is dated, it has inspired multiple adaptations. It's been on film, in theater, and also lampooned in television shows and films (ranging from *The Transformers* to *Black Knight*). I first heard of the book while reading the Carl Sagan novel *Contact*, which used this key line a couple of times: "'Bridgeport?' Said I. 'Camelot,' Said he." It was meant to represent the bridge between different ages in the book, when there was confusion about how Bridgeport could appear like Camelot to our traveler's eyes. Sagan also used this confusion as a metaphor between civilizations, including alien civilizations, which are vastly separated in both technology, and the way that their societies think about ethics and philosophy.

In 1989, the magazine *American Heritage* wrote that the book is one of the "more disturbing exercises of the American literary imagination, a brilliant comic fantasy that turns savage and shakes itself to pieces." Initially it's a simple satire of Mallory's work, but pretty soon it descends into a commentary on the progress of the modern age that doesn't quite hold up. "For those at the bottom of the economic order, the 'law of competition' had merely replaced chattel slavery with wage slavery," the magazine writes.[3] "Edward Bellamy's novel *Looking Backward*, published a year before Mark Twain's, compared American society with the Black Hole of Calcutta— 'its press of maddened men tearing and trampling one another in the struggle to win a place at the breathing holes.'" And that was indeed the legacy of the early Industrial Age. People used to live in the farmlands, and now had to migrate long distances to the city. Work was menial and took place under poor conditions, leading to illnesses and deaths in many cases. Cities were far

---

3 Kaplan, Justin. (1989, November.) "A Connecticut Yankee In Hell." *American Heritage.* Retrieved from https://www.americanheritage.com/connecticut-yankee-hell

from clean, with many jurisdictions simply throwing slop into the streets, or dealing with quality issues in the local water as microbes spread. And until the introduction of unions in the early twentieth century or so, most workers worked extremely long days with little breaks—that is, most except the owners, who received profit from a lot of other people's work.

Are things any better today, you may ask? Union jobs are diminishing, steady work is harder to get, and most of us are free agents whether we like it or not. Health coverage is tricky if you're a freelancer. Money is harder to get, as it arrives in spurts, if at all, and you often have to take work for longer hours or under worse conditions than somebody in a workplace. And like the environment of the early twentieth century, this huge group of freelance workers are poorly covered by workplace laws, forcing people to rely on the expensive legal system when things go wrong. There are no easy answers, but clearly people are struggling and there must be a massive economic solution to help those in need.

ARMAGEDDON 2419 A.D. (1928)

There was once a time when people received their science fiction stories in magazines delivered to the door. One of the many tales out of this time was *Armageddon 2419 A.D.* which was the inspiration for the more popular *Buck Rogers*. Here's how the story unfolds.

The Science of Time Travel

This story has some racial undertones that are a little startling to consider in the twenty-first century, so bear that in mind as we go through the tale. Essentially, it's the twenty-fifth century in the United States and the country is under occupation by the Hans (an East Asian ethnic group) that took over the country in 2109 AD due to a bit of alternate history. After the events of World War I, the Europeans, instead of remaining allies with the United States, decide to fight against them. The two sides are left so vulnerable by the conflict that the Soviet Union (which in real life, did have a beef with the Western world) takes advantage of the opportunity. Through a series of conflicts, the Soviet Union pairs up with the Hans and the Hans eventually take over all of North America. The Americans are left struggling, living in the forests while the Hans create huge mega-cities on their land.

So, where does time travel come in? Well, we find out about the situation through a fellow called Anthony Rogers (yes, the predecessor to Buck Rogers). He was a World War I veteran working for a company called the American Radioactive Gas Corp. in 1927. One day, while exploring an abandoned coal mine in Pennsylvania, he gets trapped in a cave-in. But there's something special about this cave, according to the story. Due to the radioactive gas in the region, he becomes stuck in a state of suspended hibernation and remains there for 492 years. When he wakes up, after some time wandering around the forest, he eventually finds the sheltered Americans and works with them to fight against the Hans. Quite obviously, radioactivity tends to kill people (unless you're a superhero of some sort, in which you get a genetic mutation and become very powerful) and its relation with time travel is, to say the least, a little sketchy. But the practice of suspended animation has been a common trope of time travel stories, including our old *Rip Van Winkle* tale earlier in this chapter on page 9.

Suspended animation, however, sounds a little bit like hibernating. It's something that some animals do during the wintertime to reduce the need to consume energy at a time when food is not as abundant. As How Stuff Works explains, "Physical activities like walking and breathing burn energy. Pumping blood and digesting food burns energy. Even thinking burns energy. For warm-blooded animals, a lot of energy is burned just keeping our body temperature where we need it. Even when we're sleeping, we're burning energy."[1] During hibernation (which, funny enough, How It Works completely relates to suspended animation), the animals can enter a state where their energy needs are less. They reduce their heart rates and breathing and their body temperature diminishes. Bodily functions such as eating and going to the bathroom also cease. Animals sort of instinctively know when to get into this state, depending on factors such as the temperature, the amount of food they have on hand, the amount of light visible outside, or even just plain old rhythms of the seasons. Could humans access this state of hibernation? Perhaps we could, if we could figure out how to hack the endocrine system (the glands in our body that affect

---

1   Grabianowski, Ed. (n.d.) "How Hibernation Works." How Stuff Works. Retrieved from https://animals.howstuffworks.com/animal-facts/hibernation.htm

the hormones being released). It's at this level that we can gain control of our metabolism (through the thyroid), our fat reserves (pituitary), and the amount of sugar that is required (insulin), among other factors.

In 2017, I spoke to a company called SpaceWorks that is examining the possibilities of hibernation in humans during long space voyages. This would not only reduce what a ship needs to carry (in terms of food and water and oxygen and the like) for these passengers to survive, but it would also get over the problems of boredom during trips to Mars (which could take several months under current technology). It may even allow the ability to extend a human's lifespan, although bear in mind that's all theoretical.

SpaceWorks received some funding from the NASA Innovative Advanced Concepts program, which examines very nascent ideas and tries to bring them to a more advanced state of technology. SpaceWorks, in 2017, was considering using "mild hypothermia therapy" as a roadmap, because studies have shown that with even a small reduction in the body's core temperature, the metabolic rate in humans does diminish. Does this sound like early-stage research? Yes, very much so. We understand very little about how the human body works, and it will take time to do the proper research and to also make sure that it is safe for people. There's a large regulatory process for manufacturing pharmaceutical drugs that demands steps such as controlled trials, which take years to run through and to validate through peer review. Altering metabolic activity would surely be even more complex.

You may think that hibernation is "cheating" in terms of time travel possibilities, but it's actually the most accessible way that we have right now of going into the future. And it is a viable plot point for so many science fiction stories, ranging from *Rip Van Winkle* to *Idiocracy*. But whether it's safe, and whether it's possible, will likely take decades of research to figure out.

The Science of Space Travel

One of the enduring legacies of the *Buck Rogers* franchise was to make people more familiar with the idea of space travel. Remember, *Armageddon 2419 A.D.* was written in the late 1920s, not only when space travel wasn't a thing, but when rockets were still a technology that reeked of so-new-we-can't-rely-on-them. Even fifteen years later, when the Nazis were lobbing rockets at London from Germany, spanning the short distance of a few hundred miles was tricky because these missiles had very early-stage navigation systems. Also, many rockets didn't even make it off the ground, exploding on the pad.

Instead of going into the same story about the space race that most journalists talk about, here I'd like to focus on some of the challenges of traveling for a long while in space. I mean, specifically for humans. Robots are, of course, prone to problems like running out of fuel or becoming less effective due to radiation, but humans have a host of different needs—oxygen, water, food, and not getting bored or overworked. (Lest you think that being in space is "enough" to get over the psychological problems, just ask the Skylab 4 crew in 1973, who decided to slow down their work because of communication issues between themselves and Mission Control. That's just one example of miscommunications between Earth and space causing psychological problems.)

Today, NASA uses the International Space Station as a platform to investigate what happens to the human body after a few months (or even a year!) in space. It's slow work because the space station can only hold so many people at a time, and the current capacity to bring up crews is only three people per Russian Soyuz ship, four times a year. By the time this book is published, however, the crew capacity may increase as commercial crew vehicles in the United States become available.

Scientists are engaged in a host of experiments for astronauts to perform, examining how their body reacts. In general, we know that despite astronauts exercising for about 90 minutes a day, their bones tend to weaken, their muscles diminish, and their fluids shift. After six months in space, a typical astronaut returning to Earth has trouble walking and feels very dizzy while standing up, with symptoms typical to a senior. There is emerging genetic research on astronauts as well, most prominently discussed with the "twins study" that took place while Scott Kelly was in space from 2015–2016, and his brother Mark (who was also an astronaut) remained on Earth. The researchers tracked some changes in gene expression, which means changes in which genes are activated to make proteins, in response to the environment, according to Space. com.[2] Genes were being activated, scientists said, in response to functions such as activating the repair of DNA, regulating the immune system, and changing the length of telomeres that protect the ends of chromosomes.

Why is all this activity taking place? This is all to protect our body from the rigors of spaceflight, which include dangers such as increased radiation and floating around all the time (which at first sounds relaxing, but is also tough on the body because it is used to operating in gravity). Scientists are still trying to understand all the changes that take place in space, and one thing that is particularly worrying them is eyesight in astronauts. Many astronauts come back from space with permanently changed eyesight, and the causes are still poorly understood. It could be anything from the fact that people are floating, to something that is unique to the space station environment—but you can be sure that NASA will want to investigate this thoroughly before sending somebody to Mars.

At the time that I'm writing this paragraph, NASA has been operating the International Space Station for twenty years and

---

2 Lewin, Sarah. (2019, April 11). "Landmark NASA Twins Study Reveals Space Travel's Effects on the Human Body." Space.com. Retrieved from https://www.space.com/nasa -twins-study-kelly-astronauts-results.html

is preparing to move people on to other destinations such as the moon (they are aiming for a moon landing by 2024) and then on to Mars. Operating farther out in space will be even more demanding for astronauts, because being on the moon will expose them to space radiation and abrasive dust, while being on Mars will be such a long journey from home that there will be a delay in space communications between planets.

NASA wants to learn as much as possible about the human body in space before venturing farther out into the universe. And by the way, that holds for other countries as well that offer astronaut programs today, or plan to offer them in the future. It may be that another international coalition goes to the moon, or that certain countries will want to go by themselves. The ultimate challenge will be figuring out how to sustainably move across the universe.

The Legacy of *Buck Rogers*

For a couple of decades, *Buck Rogers* was how a generation learned about space travel. There were jokes about rockets and spacesuits and other things looking like Buck Rogers when the space program started up in earnest in the late 1950s and early 1960s. While this generational influence has, of course, faded over the decades, people who are old enough to remember reading *Buck Rogers* as children still would appreciate the reference. If you ever watch the 1983 movie *The Right Stuff* (that covers the early space program), you'll see some Buck Rogers jokes in action. For example, a famous line in the movie says "no bucks, no Buck Rogers," meaning that the space program will need to receive continued government funding and support to let astronauts fly. (Who knows, with the rise of private space travel, this argument might be fully irrelevant in another decade or so.)

Over the decades, *Buck Rogers* inspired numerous adaptations in radio, film, television, movies, toys and even a world's fair. The estate also authorized sequels to *Armageddon 2419 A.D.* in the

1980s. While it has been changed a bit over the years, some critics of today point out that the original franchise had racial undertones (which, to be fair, other stories of the era also struggled with). Buck Rogers no longer seems like a fantastical thing because we live in the era of Buck Rogers today. We have had rotating crews of people living continuously in space for twenty years. We send stuff into space several times a month, and with the rise of microsatellites and private companies, that's sure to accelerate to several times a week. NASA and other agencies are talking about moving crews across the solar system to the moon and Mars, and robotic spacecrafts have already explored every planet and many of the moons.

Perhaps this is another reason that we no longer refer as much to the *Buck Rogers* franchise, but still, it had an indelible effect on the people who grew up with it.

LEST DARKNESS FALL (1939)

Written on the eve of the Second World War, *Lest Darkness Fall* is an early example of "alternate history," which is to say history that never actually happened, but which takes place in fiction. This book has influenced authors such as Harry Turtledove to do their own works in the coming decades.

The Science of Time Travel

The time travel in this story takes place when American archaeologist Martin Padway goes to Rome and takes a look around the Pantheon, which was a temple in early Roman times and later converted to a church. While he is there in 1938, a thunderstorm comes up and the area gets struck by lightning. Padway then finds himself in Rome in the year 535 AD. This is an interesting time in the history of Rome. Taking place long after the time of Plato and Aristotle and Caesar, in the sixth century Italy's ruling group was the Ostrogoths. They actually only ruled Rome for about sixty years, in between the end of the Western Roman Empire (the classical Roman empire to which I referred) and the Byzantines. The year 535 AD is a key moment in this turnover to the Byzantines because it's the beginning of what was later called the Gothic War. Rome was attacked multiple times and even though the Byzantines "won," it was too hard for them to hold on for long. (In fact, some historians think the decline of the Byzantines was the beginning of the Dark Ages in Europe, but that's a discussion for another book.)

Like the protagonist in *A Connecticut Yankee in King Arthur's Court*, Padway sets about trying to make his immediate environment more comfortable. His innovations include teaching some of the clerks (what we would consider "bookkeepers") the Arabic numerals, rather than the Roman numerals used at the time. He even puts together a printing press and sends newspapers out for reading. He's not always successful—his endeavor to make gunpowder a thing fails, unlike the Connecticut Yankee. But clearly, he's adaptable and before long, he begins to use his knowledge to create an alternate history of what happened during this turbulent time. We won't get into all the historical changes that Padway affects, because reading his work against the real events of history is most of the fun. That said, by the time he finishes his changes, it appears that Rome and Italy are well poised to continue sharing knowledge for the indefinite future, unlike in real life when we lost records of their work for many centuries. So in other words, the Dark Ages will not happen and to echo the title of the book, darkness will not fall.

How did Padway travel back in time? Well, it's using a lightning bolt, which you're going to see a few times in this book, including *Back to the Future*, which sort of parodies the genre. Lightning does carry a lot of energy and it is plausible that perhaps it would provide power for a time travel device, but that's not really made clear in the book. As one modern critic put it, "there's barely a handwave of explanation as to how Padway makes his way across time."[1] In our modern eyes this seems a little . . . disappointing. But as we've seen from looking at time travel books before this, the mechanics of time travel were not well-defined for most of the early history of the genre. Of course, 1938 is well after the Industrial Age began, but it appears the author is far more interested in ancient history than any modern technological wonder that could have brought Padway so far back in the past so quickly.

A criticism we can offer is that the coincidence seems a little too convenient. Why would a scholar of ancient Roman times be

---

1   Walton, Jo. (2008, Aug. 14). "Has Queen Amalasuntha Been Assassinated Yet? L. Sprague de Camp's Lest Darkness Fall." Tor.com. Retrieved from https://www.tor.com/2008/08/14/lest-darkness-fall/

transported back to exactly the period that they studied? While Rome was a mighty empire of a few centuries, it seems improbable that not only would the scholar go back to time to when Rome existed, but even more so to the very year that they can recite in detail. It's like a stock market historian who only studied technology stocks going back to the very day that Google made its initial public offering (on Aug. 19, 2004), rather than any of the other years in which the search engine existed. Yes, you can affect the future, but only if you know exactly what happened and at what time. This is one of the weaknesses of the alternate history genre. You need to have such a detailed knowledge of what happens in the timeline that you need to go back and change things to the very second. This is also discussed in detail in the book *11/22/63*, which sees the protagonist not only go back in time to try to stop the Kennedy assassination, but to make a sort of "practice run" with a brutal murder in the newspapers. As you'll see in that chapter, it's not so easy to change time even if you know exactly what happens and when.

The Science of the Printing Press

It's more and more likely these days that you consume your information electronically. You may not even think about that fact very much, given that we're surrounded by devices that can deliver information from around the world instantly. Amazon's "one click" will get a Kindle title delivered to a small device instantly. Or your free web browsing will instantly allow you to compare a big piece of international news—maybe the next US presidential election, for example—not only in the United States, but in China, South Africa, and France, as long as you can read the local languages or find ways to translate them. Until about the 1400s, however, mass distribution of information just wasn't possible. People lived in rural communities, and literacy was not widespread. Any books that were made had to be handwritten and hand-copied, so you can imagine how laborious a process that was. It was the invention of the printing press that changed all of that. The invention of the printing press is most usually attributed to Johannes Gutenberg in 1439, although it's possible that he might have borrowed some of the innovation from other countries. "Gutenberg's movable type involved metal letters at the end of a metal shaft that could be produced in quantity, arranged, and moved around so that mistakes could be corrected easily, and the type could economically be reused for the next project," explained the American Society of Mechanical Engineers (ASME). "Both of the earlier methods of reproduction were expensive and time-consuming. The combination of movable type, oil-based inks, and a workable press revolutionized European bookmaking and spread rapidly across the continent, and later the world."[2] The work that he is best known for, in fact, is the Gutenberg Bible. It was finished in 1455 and sold at the Frankfurt Book Fair that same year, according to ASME. (In fact, the Frankfurt Book Fair to this day is a major event for book publishers in advertising upcoming books, so talk about a powerful legacy for printing to have the same fair doing advertising for books five centuries later!)

Gutenberg's work initially focused on religious publications, in a nod to how much of an influence it had on society at that time. But over the years, the printing press has been used for many other means. Today we still use printing presses to create physical books and newspapers, although their use is in decline thanks to the rise of ebooks around the world. In speaking with one publisher in 2018, for example, I learned that Canadian publishers must pre-order their big book orders with printers because the printer needs to make sure they have enough paper on hand. Since Canada's economy was basically founded on pulp and paper, I found that fascinating to hear how much that society has changed.

2　Giges, Nancy. (2012, May 21). Johannes Gutenberg. The American Society of Mechanical Engineers. Retrieved from https://www.asme.org/topics-resources/content /johannes-gutenberg

So what did the printing press do for society? That's a whole book in itself, but these are some of the most cited effects. Literacy rose sharply, and especially since it was plausible that some people could read religious books by themselves, this led to more interpretations of what the Bible (and other religious books) said. The printing press is credited with spreading the philosophy, art, and wisdom of the Renaissance, which itself resurrected some of the classical thinking from the Romans and the Greeks. The printing press also sent information from Europe abroad to its colonies, keeping those who could read there apprised of what was going on in their ruling country. (This latter is by no means a fully benevolent fact, as first of all the Europeans could spread propaganda that way, and second of all, not everyone could read or, if they were slaves, was allowed to read.)

The Legacy of *Lest Darkness Fall*

I think the best legacy of *Lest Darkness Fall* is in its acceptance of alternate history. It's something that we take for granted in our shows today. (In fact, I remember watching a "time loop" episode of *Star Trek: Discovery* that put forward the idea of multiple alternate realities without needing to take much of a break to explain things to viewers, showing just how far we've come in less than a century.) Alternate histories allow us to wonder what might have been and how we could have changed the past to change the future. While we cannot actually do so, at least unless time travel is invented and we can get around the paradoxes of changing the past, thinking through alternate scenarios is a good life skill to help us grow as people. The next time you're facing a major decision, try thinking back to a time in the past when you were doing something similar. Ask yourself what you learned from the experience and if you could do better. This mental form of "time travel" is sure to help you make clearer decisions for yourself, and for the people immediately surrounding you, in the future.

PART II
THE
GOLDEN AGE

PEBBLE IN THE SKY (1950) & THE END OF ETERNITY (1955)

Isaac Asimov was one of those authors who defies genres. Even though he is best known for science fiction work, he also wrote nonfiction books as well as fantasy and mystery. His prolific legacy (he wrote something like 500 books in his lifetime) means that you can spend years getting to know his work, which sounds like a rewarding quest to me. But in the meantime, we'll focus on two time-travel tales of his: *Pebble in the Sky* and *The End of Eternity*.

The Science of Time Travel

Pebble in the Sky

This story follows the adventures of Joseph Schwartz, a retired tailor who gets caught up in a nuclear laboratory accident during a stroll down a Chicago street. Rather than being vaporized on the spot, he is transported quite far into the future, 50,000 years is what one person in the novel suggests, although other Asimov works in the series said this estimate was not correct.

If you're looking for a scientific explanation of what happened, it's probably pretty lacking here. Nuclear radiation doesn't time-travel people, but instead causes serious harm, as we know from the Chernobyl accident in 1986, among other things. That said, a time travel machine could be powered by nuclear fusion or fission in the future. That's a plausible explanation, since nuclear power provides an immense amount of power to accomplish certain things. Fusion occurs when the nuclei (or the cores) of light atoms are smashed together, generating different particles and nuclei and a lot of energy, which is the same process that happens in our sun and other stars. The sun is in a slow process right now of converting hydrogen to helium. In a few billion years, when that process finishes, the sun will swell up and turn into a red giant star. Scientists do hope to produce nuclear fusion reactors in the future, but the technology (as it stands now) does not easily produce more energy than is put into it. Fission is a process when the atom's nucleus is split into two or more nuclei. This was the type of bomb used during the Second World War, when the Americans dropped bombs on Hiroshima and Nagasaki in 1945 in an effort to stop the Japanese from fighting. The efforts worked, but many thousands of people died and the zones are still slightly radioactive today. The Chernobyl nuclear fission reactor in 1986 catastrophically failed, in part because workers had fewer rods to control the reaction than what the plant required. While nuclear energy is clean, accidents such as Chernobyl show the horrible side effects if it is not managed properly.

The End of Eternity

This is a bit more scientific than the last book, as Asimov here moves into fields that we're going to be exploring in detail later in the book. Specifically, it goes over the implications of causal loops, or when an event generates another and another and it's hard to figure out where the loop begins or ends. (This is something more recently covered in the brilliant *Looper*, which we'll discuss later in the book on page 180.)

This book is actually a good meditation on the problems of power and privilege, since members of an organization called Eternity—who have the ability to change time—are trying to go back in time to create the conditions for the history that they truly want. (This is an alarming discussion given how much people already try to revise history, even without the ability to time travel.) Their argument is that this is all for human happiness, which again is a sticky trope to navigate. How can one group of people know what is best for humanity all over the world?

The energy source for the time travel here is an exploding sun, which represents the scientific know-how available to scientists at the time. Today we know that the sun does not have enough mass to explode into a supernova, whereas huge stars such as Betelgeuse can. The mechanics of how the machine works are not discussed in much detail, but *The End of Eternity* stands as a fine reflection on privilege, and perhaps Asimov hoped to work more on that theme than to get bogged down in the technical details.

The Science of Prediction

Much of Asimov's work has been hailed as prescient, because Asimov was skilled at making the future believable, and in many cases, it ended up being true (as we'll explore in the next section). So, what goes into making a good prediction? How can we do better at figuring out which technologies will be relevant to us in one decade or three? One thing to remember (which is something you'll have to think about if you invest in the stock market) is that past performance does not necessarily predict the future. That's why NASA's 1960s visions of landing people on Mars in the 1980s seem so laughable today, even though the pace of the 1960s had gone from having no people in space to landing people on the moon. That's also why computerologists are so hesitant to say that Moore's Law, the idea that computing power (or transistors per square inch on integrated circuits) will continue to double every year, as it has for much of the last fifty. It's been pretty consistent so far, but barring large technological improvements we may run into a barrier.

The *Harvard Business Review* calls this the "fallacy of extrap-olation" and suggests that one way that we can avoid this bias is to imagine multiple scenarios for the future.[1] So, if we're to take the Mars mission projections of today, it's safer to imagine both that Elon Musk will swiftly commercialize the technology and we'll be sending humans there in the 2020s, and to also imagine a world where we hit a recession (or a lack of will to explore out in the universe) and Mars exploration pushes out late into the twenty-second century, if not later. There are other ways to avoid having only one view of the future, the magazine suggested. One idea is to put in multiple inputs. You can talk about this in the technological sense (in that you create an artificial intelligence that has multiple factors and weights combining to come to a decision) or in the human resources sense (meaning that you consult people with different points of view to come to a decision, something that is a useful skill for managers, teachers, and parents).

The real difficulty of predicting the future is the X-factor, that technological discovery that eludes us now. Let's talk about the warp drives of *Star Trek*, for example. As far as we can tell from the physics of Einstein, it is impossible to travel faster than the speed of light because mass and velocity are linked; as you approach the speed of light, your mass becomes infinite. That said, there are also some quantum particles that appear to communicate with each other at a faster speed, which Einstein reportedly called "spooky action at a distance." So, maybe there's a way around it. Our current technological framework does not allow for warp drives. We certainly have more efficient ways of jetting around the universe than rocket fuel—like using gravity slingshots or electricity or the charged particles from the sun—but the tradeoff with these technologies is that we have to go about things much more *slowly* to save on fuel. There were rumors a few years ago

---

1   Wieckowski, Ania G. (2018, November and December). "Predicting the Future." *Harvard Business Review.* Retrieved from https://hbr.org/2018/11/predicting-the-future

that the EmDrive, a NASA Eagleworks advanced-propulsion project, may lay the groundwork for warp drive (as strenuously as NASA denied it). The team who initially worked on the project found small amounts of thrust during lab tests, even though the engine didn't appear to be generating any. In 2018, however, a German team also tested the EmDrive. They also discovered the thrust, but said it's coming from electromagnetic interaction and not from the drive itself. So those warp drive dreams will have to stay dreams for now.[2]

The Legacy of Isaac Asimov

In some ways, Asimov was a very accurate predictor of the future. A *Forbes* article from 2014 referenced Asimov's predictions about the year some fifty years before. Among other things, the prediction said that "robots will neither be common nor very good in 2014, but they will be in existence," mentioning items such as "robot housemaids," "robots for gardening work," " 'robot-brain' vehicles that can be set for particular destinations" (or today's GPS) and improved batteries (although he predicted batteries running off radioisotopes, which is not the technology we ended up using). Of course, even a genius such as Asimov would have had difficulty figuring out some of the advances of our day. That same article cheerfully talks about a human mission to Mars being in the works by 2014; as this is written five years later, most people acknowledge a landing on Mars would likely be (at the least) in the 2030s, and that would be only if NASA meets its stated goal to land humans on the moon in 2024 and follows the plan to move forward. He also said that humans would become a "race of machine tenders" and that schools should educate its children to occupy this

---

2   Wall, Mike. (2018, May 23). "'Impossible' EmDrive Space Thruster May Really Be Impossible." Space.com. Retrieved from https://www.space.com/40682-em-drive -impossible-space-thruster-test.html

function—which obviously has a lot of problems for educational thinking, as well as society.[3]

In the immediate aftermath of Asimov's 1992 death, the *Los Angeles Times* interviewed several scientists who say they were inspired by Asimov (and in some cases, such as artificial intelligence researcher Marvin Minsky, they actually got to know him). "I remember reading the first robot stories and deciding I was going to build them," said Minsky, who was sixty-two years old at the time. "His stories asked how you could possibly impart common sense to a machine, which is something I have spent my life at." (Minsky died in 2016 at age 88.)[4] In the same obituary, science fiction writer Ben Bova said Asimov's genius was to "take any subject under the sun and write about it so clearly and so simply that anybody who could read could understand it." I would argue that even goes for Asimov's writing about the technologies that haven't been invented yet. While time travel hasn't yet happened (at least in the sense of somebody using a machine to travel to the future or the past), Asimov's stories make us believe that it happened.

Another thing that Asimov's work previewed was the current spate of "hard science fiction" stories in our media, meaning the science fiction that draws a lot upon real life to make the story work. One of the more recent spectacular examples of hard science fiction was *The Martian*, an Andy Weir novel of 2011 that turned into a popular Hollywood film in 2015 starring Matt Damon. While the premise was clearly fictional—discussing the adventures of a crew on Mars—the story used real-life orbital mechanics calculations,

---

3   Sardana, Sanjeev and Sandeep. (2014, Jan. 15). "Asimov's Vision Of 2014 And A Legacy Of Looking Forward." *Forbes*. Retrieved from https://www.forbes.com/sites/sanjeevsardana/2014/01/15/asimovs-vision-of-2014-and-a-legacy-of-looking-forward/#617a9cc732ec

4   Teitelbaum, Sheldon. (1992, April 8). "Scientists Say Asimov Put the Stars in Their Eyes." *Los Angeles Times*. Retrieved from https://www.latimes.com/archives/la-xpm-1992-04-08-vw-636-story.html

referenced several NASA missions, and used rocket and spacecraft technology that is plausible enough that it could be used in the near-future (even though these haven't quite been invented yet). And it was a darn good story nonetheless.

Science fiction encourages us to think critically about the world around us, and to look at the past and the future with an eye to what technology used to work—and what technology will work in the future. Time travel isn't something that we use every day, but if we ever end up doing so, you can be sure that Asimov's work will be reverently referenced by those future travelers.

A SOUND OF THUNDER
(1952)

Historians of science fiction will really appreciate how prescient the short story "A Sound of Thunder" is to more modern stories. Not only does it anticipate some of the paradoxes and other questions of time travel today, but the plot sounds a little reminiscent of—of all things—the Jurassic Park series. Now if only Spielberg and the rest of the crew could integrate time travel with dinosaurs, they'd really give a shot in the arm to the old franchise. But yes, this summary is about the Ray Bradbury story, so let's dive right in.

The Science of Time Travel

Here we see time travel being commercialized, which is an interesting thing (and makes me wonder what sorts of horrible commercials could be generated from this new market). Anyway, it's the year 2055 and there is a company called Time Safari Inc. that positions itself as an experiential experience for wealthy people. While today's super-rich dream about flying into space, this story suggests that the super-rich may want to indulge another fantasy—killing dinosaurs, sixty-six million years ago.

Of course, many precautions are taken to make sure that the hunting parties have minimal effects on the future as they seek out the dangerous and famous Tyrannosaurus Rex. For one thing, the animal being hunted is a doomed animal anyway—scouts go in ahead of time and (somehow) follow these animals until they track down a few that are killed or die under their watch. Hunters are also told to stick closely to a levitating path, just in case they run into other creatures or somehow (I suppose) step on a plant or something that could be a distant and important ancestor of today's life.

Predictably, one of the hunts does not go as planned and we then get to see some time ripples in the future. And really, the story does make you think critically about one modern offshoot of this practice—the people who pay a lot of money to go to places like South Africa and hunt down lions. A few weeks before I wrote this paragraph, a couple illegally killed a mountain lion in Yellowstone Park and, in a classic blunder of our modern age, decided to post about it on social media. In fact, it wasn't just *one* social media platform, but they triple-sealed their mistake and put the kill on Facebook, Snapchat, and Instagram.[1] (Which makes me wonder how we'd possibly control dinosaur-hunters, but that's a discussion for another day.)

Anyway, what are the mechanics of these people traveling back through time? Bradbury's short story leaves tantalizingly few details. It mentions a "Machine" with "silver metal" and a "roaring light." It appears that Bradbury didn't feel it was necessary to discuss the mechanics of how the Machine made it back, but simply to point to the implications. While we can't say much about the mechanical aspects of time travel in this story, we can (with apologies for a spoiler) point to how this little story kicked off a phrase known as "the butterfly effect." It turns out that one of the hunters returns to the present with a butterfly, "very beautiful and very dead," that he accidentally stepped on while wandering in the Cretaceous woods. It's implied that this one small act—which seems to be nothing, at first glance—greatly changes the future.

Bradbury's story didn't coin the phrase "the butterfly effect," but it has been noted as the first known instance of a butterfly's fate deciding the future. Rather, the phrase is more closely associated with Edward Lorenz, an American meteorologist and mathematician. Lorenz's observations date from 1961, when he was running a

1   Koshmrl, Mike. (2019, June 15). "Social media did in Yellowstone lion poachers." AP News. Retrieved from https://www.apnews.com/369e47c35a964015a8a1c06f1f5c31e7

computer model for a weather prediction. Computers in the 1960s needed a lot of manual input, and operators often worked from printouts before punching the required numbers into the computer. Lorenz typed in a number that said 0.506, a rounded number from the value 0.506127. You would think that at that level of precision, there wouldn't be much change, but Lorenz discovered that the weather prediction altered completely as a result of the rounding.[2] In a later work, Lorenz observed: "One meteorologist remarked that if the theory were correct, one flap of a sea gull's wings would be enough to alter the course of the weather forever. The controversy has not yet been settled, but the most recent evidence seems to favor the sea gulls."[3] While the original reference appears to have been sea gulls, in other works Lorenz used the butterfly, perhaps because the metaphor is prettier.

While many people might believe that a simple flap of a butterfly's wing could change the future, it's hard to pinpoint chaos theory—the underpinning of this butterfly metaphor—with that much precision. Nature is complex. We're seeing this right now with the warming of the world due to climate change. One thing that has been difficult to model is the effect of the melting of the glaciers. Glaciers, among other things, reflect the sun's heat back into space. When they disappear, they leave blacker ocean or terrain behind and the sun's heat is better absorbed, warming up the Earth. Some scientists worry about an accelerator effect once the ice is all gone, but it will probably only be a few decades before we really get to see the effects.

On the flip side, the butterfly effect shows that even a small effort might be able to change your own life. In his book *Atomic Habits*, James Clear suggests that even if you make a small effort

2   Gleick, James. (1987). Chaos: Making a New Science. Viking. p. 16
3   Lorenz, Edward N. (1963). "The Predictability of Hydrodynamic Flow." Transactions of the New York Academy of Sciences 25(4), 409–432.

at doing something useful—exercising, studying, flossing, the like—it will compound over time and turn into something larger. It's perhaps an explanation that relies much on privilege, but nevertheless, it does give us an aiming point if we're trying to improve the lives of ourselves or others.

The Science of the Cold War

It's interesting to consider the time in which "A Sound of Thunder" was written. At the time, United States and Soviet relations were taking a bad turn. These two countries had fought as allies in the Second World War, but over time, political differences were making the Soviets turn inward and pursue authoritarianism and communism, while the United States was positioning itself as a world leader and trying to meddle in the affairs of more countries with little direct lived experience. Of course, these are simplistic views of the countries, but the point is that they had their differences. The United States tried to portray itself as a bastion of democracy, and the Soviets tried to portray themselves as being strong through unification. And one of the most famous ways that this played out was in the Space Race, which happened a little after the book was published (starting in the late 1950s) but whose groundwork was already in place before. Following the Second World War, both the Soviets and the United States captured German rocket scientists and brought them into their respective militaries. Starting in 1957, the Soviet Union launched the satellite Sputnik into space and the United States caught up a few months later with its own launch, Explorer. The two sides were in a seesaw for years after that, trying to one-up each other with space-based milestones: first person, first multiple-person crew, first woman (actually, the Soviets had the only woman in space for two decades), all the way up to the first moon landing. Both sides also built up nuclear armaments and squabbled about how to legislate them.

We clearly live in a different world today, spurred in part by the Soviet Union breaking up into several countries (including Russia) and the United States inviting the Russians to participate in big projects, namely the International Space Station. The world is no longer poised between two superpowers, but is host to a rising group of nation-states that are capable of, and insist upon, forging their autonomous way.

This is a theme that we'll see in time-travel stories over and over again, that the narrator will go back to the past, something large will change, they will return to the present to discover that the "present" (at least, the present that they now live in) is changed. This theme is in fact so prevalent in time-travel literature that the series *Back to the Future* (which you will see in Part III on page 109) lampoons the genre. But it is fun to think about how history would change if you went back in time. In fact, the Stephen King novel *11/22/63* talks about an attempt to help John F. Kennedy survive his assassination attempt, and even for such a gesture of hope, it may not be as simple an idea as it would seem.

The Legacy of "A Sound of Thunder"

Bradbury is such an influential author that when people think back to "A Sound of Thunder" they often relate it to the rest of Bradbury's work in general. Bradbury was one of those people who brought science fiction into the mainstream, which makes him an early predecessor of all of those superhero and dystopian movies that are so popular in theaters today.

Bradbury was preoccupied with the Cold War, and that—along with other easy-to-understand themes—makes his work a staple in middle schools, wrote the *New York Times* upon his death in 2012.[4]

---

4   Kakutani, Michiko. (2012, June 6). "Up From the Depths of Pulp and Into the Mainstream." *The New York Times*. Retrieved from https://www.nytimes.com/2012/06/07/books/ray -bradbury-who-made-science-fiction-respectable.html

He's often one of the first writers who awaken students to the enthralling possibilities of storytelling and the use of fantastical metaphors to describe everyday human life. His finest tales have become classics not only because of their accessibility, but also because of their exuberant 'Twilight Zone' inventiveness, their social resonance, their prescient vision of a dystopian future, which he dreamed up with astonishing ingenuity and flair.

He also had a reverence for the written word and its power upon people (something that perhaps pop culture best understands today by the feeling tattoo artists have when, to paraphrase the words of a recent contestant on *Ink Master*, they put art on somebody's body and it travels with them for the rest of their lives). In one of Bradbury's last interviews, the BBC noted that for him, it gave his words immortality—as well as other folks who had worked in novels, books, newspapers, and the like. " 'Books are alive, you see,' Bradbury said in the interview. 'They're not dead, they're alive. This book is alive,' he says, tapping his hand against the copy of *Fahrenheit 451* in front of him. 'That's me. That's my flesh.' "[5]

Now, in case you think that Bradbury would have had a similar feeling for the online news sources, social media, ebooks, and other non-printed words that we consume today, the ninety-one-year-old had definite negative feelings about that sort of a thing. The BBC records him as "slamming a palm" on the table while declaring that electronic books were junk, fake, and stupid. To understand why, it's good to briefly mention the plot of *Fahrenheit 451*, which is about a future American society that burns books, a commentary on censorship and also on a society that does not read as much literature now as it did in the past. Bradbury told

5   Novak, Matt. (2012, June 8). Ray Bradbury: The day I read to a legend. BBC. http://www
    .bbc.com/future/story/20120608-meeting-the-master-ray-bradbury

the BBC that his novel was meant to keep physical books ("real books," in his mind) in libraries forever. "I saved the libraries," he said. "With [*Fahrenheit 451*], I went out and lectured and I saved libraries all around California."

Now, you may not agree that electronic books are stupid or junk—and as a person who uses my Amazon Kindle for hours every week, I'm guilty of upsetting this master about whom I am trying to write so reverently. But Bradbury does have a point, even though ebooks weigh nothing and are easy to transport. Inevitably, ebooks are easier to lose or have deleted than the physical copy. There have been a handful of instances where the publisher recalled an electronic book, wiping it off people's devices. And I'm nervous about putting in suggested "updates" for the books, out of fear that my electronic bookmarks and highlights will be removed. So here, Bradbury makes a forceful point, even from beyond the grave. More of us should probably listen to his wisdom.

TOM'S MIDNIGHT GARDEN (1958)

Children's literature sometimes has the most profound themes, so while it may be easy to dismiss *Tom's Midnight Garden* as a kid's book, there is so much more to it. It's a book about lost youth, about wonder, and about trying to reconcile the past with the present. This science fiction book is also written by a female, Philippa Pearce; unfortunately, women (and people of other genders) are not as prevalent in science fiction as men, so let's highlight her work and see what we learn.

The Science of Time Travel

The book opens with a strangely modern premise, given the number of anti-vaccers that has led to the rise of childhood diseases such as mumps and measles. Tom Long is a young boy living in the "modern day" (the 1950s, when the book was published) whose brother, Peter, gets the measles. So Tom goes to the house of his Uncle Alan and Aunt Gwen—it's a very old house to a kid, as it was built about sixty or seventy years before the story begins. His relatives actually only occupy the upstairs flat of the house, because England (like much of the United States at that time) was undergoing explosive growth of the suburbs. The former grounds of the house are surrounded by houses. Tom only has a small area to play in, and cannot leave the house out of fear of infecting nearby children.

One night, he's lying awake very late (past midnight) and then he realizes that the grandfather clock is striking thirteen notes instead of the expected one note. This budding investigator gets up to find out what is happening, and finds out that outside the back door is a sunlit garden dating to the Victorian age (when the house was built). Outside, Tom discovers another child named Hatty, and here is where the strangeness begins. Tom goes out every night, enjoying it so much that he continues past the mandated quarantine period. But he discovers that Hatty is growing up faster than he is; she eventually gets so old that she becomes an adult and begins falling in love with an acquaintance. (Unfortunately, we need to reveal a spoiler here to fully discuss the book, so if you would like to skip to the next chapter, stop reading this paragraph right here.) At story's end, Tom goes outside past midnight and can't find Hatty or the garden anywhere. He begins calling Hatty's name as he crashes into items in the yard, causing a commotion. His uncle and aunt summon him the following morning to apologize to a neighbor, Mrs. Bartholomew. It turns out that *she* was Hatty and had been dreaming every night of her childhood. The very last night, there was no garden because she dreamed of her wedding to "Barty" or Mr. Bartholomew.

There are a couple of interpretations that you can have for this "time travel" story, the first being that you want to dismiss it because it is a projection of an older person's mind, and the second being that there is a lot of wisdom in it because as we get older, we can time travel into our past through thinking of memories. I think there is value in both of those interpretations. From Tom's perspective, he is time traveling (although how a young boy is hooking into an older woman's dreams is a little bit beyond scientific explanation). It's implied that he's not appearing in the same spot over and over again with a woman zipping around the timeline, because the garden also changes when Tom goes inside of it. We'll see some debates in this book about whether you can travel in time *and* space at the same time, and this book is firmly in the "time only" camp. Because when Mrs. Bartholomew dreams of her wedding, Tom cannot access it since it is outside the garden.

And is Mrs. Bartholomew time traveling? I suppose that's whether you count dreams as time travel. It's a common trope, for

THE GOLDEN AGE   61

example, for us to travel back in time in dreams to situations in elementary or high school—forgetting the way to an exam, losing stuff in a locker, that sort of thing. Luckily, Mrs. Bartholomew is remembering happy times in her dreams, but all the same, you have to decide here whether dreaming or recalling something is true time travel, in the sense that memories are time travel. I think they can be, but not everybody may agree with me. Mental time travel (or chromesthesia), incidentally, is a scientific term. "Mental time travel" was first used as a term by Thomas Suddendorf and Michael Corballis in 1997, while chromesthesia's term was created by Endel Tulving.

These are obviously long after the events of the story of Tom's Midnight Garden, but it does show a measure of scientific respect for how memories and time travel may be interlinked. And there are a lot of questions associated with this scientific research. Is this ability unique to humans, or can animals also freely engage? How well do children do this as opposed to adults? When we see activity in the brain associated with memories, is that time travel? When somebody has a disease such as dementia, how does that affect their ability to mentally travel through time? These questions unlock one thing, at the very least—we need to understand more about the brain, and how it stores memories and how it affects beings (including human beings). By doing so, we'll better be able to construct what it means to be yourself, because memories (inevitably) form a large part of who you are. Perhaps Mrs. Bartholomew's dreams of her wedding night symbolized how important her marriage was to her. And for you, perhaps the dreams you have about your family and friends help cement those relationships in real life.

Pearce herself used a time theory put forward in a 1927 book called *An Experiment with Time*, written by J. W. Dunne. While the book was never accepted by mainstream science (its premise is too hard to prove), Dunne explored precognitive dreams, or dreams that foretell the future. (We all have some of those on occasion, but we can simply argue that what happens out of those dreams can be a coincidence.)

There and Back Again

I've explored a bit of the science behind themes of other books, but the ones of *Tom's Midnight Garden* are more about philosophy and life than about science—you know, the usual truisms about aging and friendship and finding value in everyday life. So rather than being scientific, I thought that instead I could offer a comparison with a book of vast popular culture that everyone is familiar with: *The Lord of the Rings* by J. R. R. Tolkien (who grew up in the Victorian Age). There are many, many themes in *The Lord of the Rings*, but there is an inherent respect for nature. Sam is of all things, a gardener, and he finds those skills useful when working with his bestie, Frodo, as they journey to destroy the ring. Sam displays patience, creativity, and even a love of the elves (creatures very closely bound to nature themselves), and time and again, it helps. Both the book and the movie versions, for example, show an incident where Sam trusts his weight to an elvish rope while going down a steep cliff, and then manages to unravel the still-tethered rope from the bottom of the cliff. But *The Lord of the Rings* theme that is even more relevant to *Tom's Midnight Garden* is its "there and back again" nature. After the Hobbits finish their quest, they return back to the Shire; in the book, there's even an entire war that is fought on the Shire soil. The complaint from many readers is that the story has "too many endings" and it should have stopped sometime before the Hobbits get back, or immediately afterward.

In the garden of the Shire, the Hobbits use the skills that they have implemented in their everyday lives. That's something that Tom eventually does as well. In the garden, he learns about how to love somebody unselfishly, regardless of age or gender. It sounds like a small skill, but think about the number of politicians or people in power who are actively campaigning against homosexual rights, or passing laws restricting the ability of students to pay back loans. People seem to forget that we all came from somewhere and that we need a lot of help to get to the culmination of our career; worse, there are some folks who have less privilege and need an extra helping hand. So, I hope Tom went on in his (fictional) adulthood and kept his eyes open for people who needed allies.

The Legacy of *Tom's Midnight Garden*

Tom's Midnight Garden is not a book that you'll usually see in time-travel analyses. Despite being adapted for television in the United Kingdom, it's also not one of those books that has penetrated the mass popular culture, at least yet. But at least some fantasy authors today have been influenced by her work. Philip Pullman, the author of *The Golden Compass*, among many other works, was voted in 2007 as the best winner of the annual Carnegie medal (for children's literature) published in the last seventy years. While Pullman easily won with 40 percent of the vote, and Pearce was the follow-up with 16 percent, Pullman told the *Guardian* that he felt Pearce was more deserving of the honor. "Personally, I feel they got the initials right, but not the name," he said. "I don't know if the result would be the same in 100 years' time; maybe Philippa Pearce would win then."[1]

Perhaps it's best to think of Pearce in the context of other children's literature about magical gardens. I'm thinking specifically here about C. S. Lewis's land of Narnia and Frances Hodgson Burnett's *The Secret Garden*, which both are written in (or evoke) the Victorian age. Gardens were a popular facet of the high-class houses that these books discuss, and it is there that children find a true place to be themselves. The Victorians were known for their love of order and of rules, and children were sometimes seen as

---

1 Ezard, John. (2007, June 22). Pullman children's book voted best in 70 years. *The Guardian*. Retrieved from https://www.theguardian.com/uk/2007/jun/22/books.booksnews

hindrances. In the garden, or outside, they could run around and be more of themselves. Gardens are also where these children forge their independence. Tom learns to trust Hatty, rather than seeing her as an annoying kid. The protagonist of *The Secret Garden* works hard to transform the dying garden into something worthy of its past, and along the way learns resilience in herself. And Narnia, which is about an entire magical land that includes beautiful gardens, is a place where children can actually rule the kingdom, and rule it with peace. The adults are portrayed as evil, or conniving, or not understanding what the people really want, presumably because they got too old to keep their eyes open.

While you can argue that *Tom's Midnight Garden* is a book rooted in nostalgia and longing for the past, I think the greater lesson here is that there are even small ways in our world that we can make an impact. For Tom, he goes into a garden and makes an unexpected discovery, forever changing the life of Mrs. Bartholomew and himself. It's a small discovery, one between two people, but in a human life, being able to help at least one other person is a good thing to aspire to. Not all of us have the power, or the riches, or the time, to devote to big causes. But we can certainly think about ways of tending our own backyards, either metaphorically or physically. Think about one person with whom you can make a difference, or one thing that you think you can change on your own, and focus on that person or thing. Meaning in life is partly found from helping others. And I wish you luck in finding your way of doing so.

DOCTOR WHO (1963-PRESENT)

Doctor Who literally has a device embedded in it that spans time and space, so it seems worthwhile to do a deep-dive into the **Time And Relative Dimension In Space (TARDIS)** device that sparks so many of the Doctor's adventures. How does it work, beyond being bigger on the inside than on the outside? Let's take a look, because believe it or not, scientists have already looked into the issue.

The TARDIS

To understand the *Doctor Who* franchise, one of the basics you must know about is the TARDIS. It's cleverly designed as a British blue police box (where calls for help could be made) from the 1960s. Luckily, while most of these were removed in the 1970s, a handful persist today and the cultural relevance remains. The TARDIS has the freedom to move across both time and space, allowing characters to meet all sorts of historical figures, including Agatha Christie, William Shakespeare, Adolf Hitler, Charles Dickens, and Richard I of England, among many others. It also allows the characters to witness events in history (or in the future) or even travel to other planets, such as the elegant episode "The Waters of Mars," which showed an ethical dilemma that the Tenth Doctor wrestled with when he learned a secret of a Mars colony.

Let's examine the TARDIS scientifically, shall we? Physicists Ben Tippett and Dave Tsang wrote an awesome paper in 2013 called "Traversable Achronal Retrograde Domains in Spacetime" (which yes, spells TARDIS as well).[1] The paper was released in

---

1   Tippett, Benjamin K. and Tsang, David. (2013, Oct. 29). Traversable Achronal Retrograde Domains In Spacetime. Arxiv.org. Retrieved from https://arxiv.org/abs/1310.7985

pre-published format on Arxiv and later appeared in publication as an article in *Classical and Quantum Gravity* in October 2013. Here's how the argument of the paper goes: the physicists hearken back to Einstein's theory of general relativity, that shows space and time are intertwined. Einstein's theories are where you begin to run into fun paradoxes, such as a twin traveling around the speed of light, returning to Earth to meet their twin, and discovering that the twin on Earth aged faster than the twin in space. This is a real thing, even though it sounds counterintuitive; time slows down for the observer going nearly the speed of light. Large objects in space (such as galaxies and stars) warp the fabric of space-time around them, and around black holes the warping is extreme. It is possible, we think—although we can't quite get it to work in the real world yet—that we could fold back one of those space-time warps and create a time loop, or more properly speaking, a closed time-like curve.

Now, we need to talk a bit about light and how it works. Light has a finite speed in space and it appears, as far as we can tell, that nothing can go faster than it. If you can imagine an explosion of light in space—maybe a star blowing up—light can only travel so far from the explosion in a certain amount of time. And light will eventually become so dispersed as it travels that it will become invisible to outside observers. This area of light visibility is called a "light cone" in the paper. Therefore, the only way you can break out of the light cone is to go faster than light itself—another fun paradox. How would we do that? By exploiting the shape of the closed time-like curve. The mathematics show that the light cones are all tipped slightly and that it's possible to jump from one to the next—you'll have to trust me on that one, as even my math skills aren't good enough to explain exactly why. The bottom line is, that's how the TARDIS works, these mathematicians argue, by leaping between cones and therefore breaching the boundaries of space-time.

Business Insider has another fun tidbit about the TARDIS. In episodes where the machine is traveling into the future, the opening credits show a space-time vortex in red. "Fans speculate that's because light emitted by an object moving away from a viewer (perhaps into the future) shifts toward longer wavelengths . . . as the object moves away, the wavelengths of its light stretch out, making it appear more red," explains Kelly Dickerson.[2] TARDIS properties have actually produced some copyright issues here on Earth between science fiction and real life. "In 1996, BBC applied for a trademark on the blue police box to take advantage of a merchandising void that had existed for quite some time. Naturally, the Metro Police objected to this trademark," ScreenRant explained.[3] "Given the sheer volume and wide array of TARDIS merchandise that has flooded both the UK and US markets, we'll give you one guess as to who won this patent war," it continued. "In 2002, the Patent office ruled in favor of the BBC, giving it exclusive rights of the blue police box design, and giving the people the freedom to buy TARDIS Chia Pets and TARDIS Snuggies. What a world we live in!"

Everyone Is Important

If I may move here into the newer series of *Doctor Who*—apologies to the classicists, but I am trying to speak to Generation Z here—it strikes me that one of the overall themes of the newer doctors is acceptance. The doctor treats everybody that he (or she) meets with the same kind of reverence, trying (like Gandalf in *The Lord of the Rings*) to hold back on directly intervening in matters unless it is of

2 Dickerson, Kelly. (2015, June 9). The physics of Doctor Who's awesome time-traveling ship aren't exactly science fiction. Business Insider. Retrieved from https://www.businessinsider.com/physics-of-the-doctor-who-tardis-box-2015-6

3 Easterbrook, George. (2016, Nov. 27). Doctor Who: 15 Things You Didn't Know About The TARDIS. ScreenRant. Retrieved from https://screenrant.com/doctor-who-tardis-facts-trivia/

supreme importance to the universe. The Doctor listens to all sides and considers all points of view, and the show even covers touchy subjects that are still thorny to discuss today. In 2018, for example, the *Guardian* pointed to recent episodes where The Doctor had episodes about Rosa Parks, the woman of color who refused to sit at the back of the bus where people of her race were supposed to be, and the 1947 partition of India, which touched on the many tangled threads of colonialism and independence.

But there's a criticism lurking behind this push for equality, the *Guardian* adds: "Incredibly, with nearly 300 Doctor Who stories having made it to screen over the last 55 years, acclaimed children's author Malorie Blackman is the first black writer the show has ever had, while award-winning Vinay Patel is the first writer of Asian descent to contribute a story."[4] And the series also has had criticisms about how it portrays race, particularly in the early years (when it was mostly white people) and even in some later years (such as "Turn Left," a 2008 episode that has some unfortunate connotations about Asian fortune tellers, the *Guardian* says). It's not easy to think about how to make the episodes more accepting, and certainly the writers should be applauded for their efforts. Clearly, there needs to be more efforts to involve the community whenever one is writing about a particular group (especially in sensitive matters of gender and race). And the show should also be careful not to hire writers *only because* it wants to tackle a sensitive topic. In other words, why not regularly have people of color on staff, or include a diverse mix of genders? How you grow up in the world shapes your entire perspective, not just the parts directly relating to your heritage.

Canada has recently had a reckoning (of sorts) as the public is realizing their poor treatment of indigenous peoples not only

---

4   Belam, Martin. (2018, Nov. 13). "Is Doctor Who finally getting it right on race?" *The Guardian*. Retrieved from https://www.theguardian.com/tv-and-radio/2018/nov/13/doctor-who-is-the-series-finally-getting-it-right-on-race

hundreds of years ago (in the colonial era) but also in more recent history, up to and including today. Some news outlets have since done an admirable job of not only writing about indigenous issues, but also including writers among the staff, such as the CBC, Canada's national broadcaster. Scientists are also making more efforts to bring indigenous people in as fully accredited researchers, rather than just participants, I learned at the Science Writers and Communicators of Canada conference in 2019. We still have a long way to go, and I can't even begin to imagine the issues that minorities face given my own privileged heritage (white, an English speaker in a predominantly English-speaking country, and identifying as a straight woman). But I hope I can do my small bit to be an ally. And if you're able to, try helping those in your community as well.

The Legacy of *Doctor Who*

If you're wondering just how influential *Doctor Who* is, just look at the franchises it inspired. If you time warp forward to *Bill & Ted's Excellent Adventure* on page 122, you'll see that the teenagers go through time using a phone booth, something that looks very similar to the TARDIS. The Fourth Doctor (Tom Baker) was a character on Matt Groening's *The Simpsons*, and his companion Leela inspired a character on that other Groening series that discusses time travel, *Futurama*. *Doctor Who* has inspired numerous TV series that included references to it; perhaps one of the cutest came on *My Little Pony: Friendship Is Magic*, which not only had a Tenth Doctor (David Tennant) inspired character named Dr. Hooves, but also pony versions of the Third, Fifth, and Eleventh Doctors (Jon Pertwee, Peter Davison, and Matt Smith).

The cultural influence of *Doctor Who* moves beyond the small screen. YouTube channel *Epic Rap Battles of History* includes a "Doc Brown vs Doctor Who" episode (Doc Brown being the time-traveling scientist from *Back to the Future*).[5] Chameleon Circuit is a British band whose song list is largely inspired by *Doctor Who* and they even are known for their own genre, known as Time Lord Rock or Trock. England's Chester Zoo named five baby penguins after Doctor Who characters in 2013: The Doctor, Dalek, Davros, Gallifrey, Sonic, and TARDIS.[6] I can also speak to its cultural influence through a very fortuitous coincidence. I was on a brief trip to the United Kingdom in August 2014 that included a very quick stop—something like an overnight stop—in London. While I was there, the Eleventh Doctor (Peter Capaldi) was announced on BBC One. I rushed out to one of the police booths/TARDIS in London, at Earl's Court tube station, and was delighted to see that fans had been there before me. There were free treats for the taking and numerous pictures of Capaldi pasted to the sides of the booth.

It's good to see that *Doctor Who* evolves with the times, although in some ways one wonders how it took so long. For example: the latest series is still running on television, and as of this writing the Thirteenth Doctor is a woman. While the rumor goes that women as the doctor have been discussed since the early 1980s, Jodie Whittaker (the first female doctor) wasn't announced until July 2017. Perhaps it waited that long because the series needed to justify it within the timeline? For example, the 2011 episode "The Doctor's Wife" included a mention of Time Lords with female incarnations, and there was a 2015 episode called "Hell Bent" that showed a black woman regenerated from a white man. It would be lovely for future Time Lords to be even more inclusive of other genders, so we'll keep our fingers crossed that it happens soon.

———————

5   ERB. Doc Brown vs Doctor Who. Epic Rap Battles of History. YouTube. Retrieved from https://www.youtube.com/watch?v=xDj7gvc_dsA

6   CheshireLive. (2013, May 2). Chester Zoo's new penguins named after Doctor Who characters. Retrieved from https://www.cheshire-live.co.uk/news/chester-cheshire-news /chester-zoos-new-penguins-named-5111119

Indeed, the built-in reincarnation has at least one writer think-ing that the show can always stay fresh with the arrival of new characters: "The show also benefits from a deeply-rooted awareness of its own mortality; rebirth and renewal are literally built into the Who concept via the 'regeneration' idea," said Graeme McMillan in *Time*, explaining it was "created as a particularly blunt method of dealing with the show's loss of its original star three years into its run—the recasting explained as 'Our hero doesn't die, he just gets reborn as someone else.'" "Regeneration has, through time and familiarity, started to seem curiously elegant," McMillan continued. "We expect the show to recreate itself with each new incarnation of the main character, and that in-built drive toward novelty helps it remain sharp, even as other shows trend toward repetition and boredom."[7]

7   McMillan, Graeme. (2013, March 15). Why Doctor Who is Pop Culture Sci-Fi At Its Best. *Time*. Retrieved from http://entertainment.time.com/2013/03/15/why-doctor -who-is-pop-culture-sci-fi-at-its-best/

JESSAMY (1967) & CHARLOTTE SOMETIMES (1969)

Women and people of other genders are not represented enough in time travel stories, as well as children, although there have been some wonderful efforts over the years. Two female authors of the 1960s—Barbara Sleigh and Penelope Farmer—created similar compelling stories about children experiencing "time slips" (much like what happens in *Tom's Secret Garden*) and then learning more about the past, as well as their present.

Time Travel in *Jessamy*

Jessamy is an orphaned girl who lives with two aunts, alternating between them during the school holidays and the school term. As she heads to one of her aunts during summer vacation, she discovers that one of the children is a victim of whooping cough (a common childhood disease of the 1960s). Like Tom in *Tom's Midnight Garden*, her destiny is altered by illness, although in this case Jessamy ends up going to a different house (the one of Miss Brindle, who lives in a Victorian mansion) to avoid getting sick.

Like many "time slip" stories of the day, the mechanics of Jessamy's time travel are left to the imagination. She goes back into the past simply by going into a cupboard (which, I suppose, one could say is a wormhole or some other technical thing, although that's not really explained in the book). Rather, Sleigh focuses on what she feels is the most important part of Jessamy's journey: trying to find people who really care about her. Her 1914 self is somebody surrounded by love and attention and affection, while her "present day" self is somebody who seems to be cared for, but

THE GOLDEN AGE 🌑 73

in a mostly detached way. Perhaps the most interesting reflection is what ends up happening to Jessamy at the end, when she finally achieves her dream of going to boarding school. Boarding school is most obviously a privilege for the rich, or for people who treat it like a mortgage and send their kids at great sacrifice to themselves. For Jessamy, it appears to be her only escape from a life in which she doesn't have anybody around her that really roots for her well-being. It's unclear if the boarding school ever provided her with the structure she sought, and it's unclear if hanging around kids of a certain class really exposed her to the "real world," but we can hope that she emerged from the experience feeling more stabilized and able to take on challenges she encountered.

Time Travel in *Charlotte Sometimes*

Charlotte Sometimes uses a slightly more plausible idea for time travel—the idea of going to sleep in a bed and waking up in the past, or the future, depending on which bed you're in! It's still not a very scientific explanation, although the rational mind can say that perhaps the girl was dreaming. But to say that would oversimplify a wonderful story.

Charlotte (like Jessamy, by her story's end) is at a boarding school and discovers that upon going to sleep at certain times, she "wakes up" as a girl named Clare living in the year 1918. Not only that, Clare and Charlotte are clearly separate people, as they write notes to each other in their own time realities for the other to read. But as in all double-identity stories, it's difficult to keep going without being discovered. The real test comes when Charlotte is unable (for reasons you will read in the story) to go back to her own time, and gets trapped where Clare lives. This means that she has to remain with her false ideas for longer periods of time, leading to the possibility of more complications. But in between, she gets to see some history as the First World War ends and she sees the celebrations.

Charlotte Sometimes is clearly a meditation on identity, and it also brings to mind how much we are shaped by the people living around us. It's fun to imagine, sometimes, what sort of "lived reality" we would have if we grew up in a different country, spoke a different language, or were immersed in a different culture. This is why travel, for those who can afford it, is a valuable experience, although this cultural education can also be achieved through getting to know others in your community and consuming media written from the points of view of people very different from yourself.

The Legacy of the Two Books

The books are still cited and read today, although they are not as popular as some of the other time travel stories that we explore in this book. *Charlotte Sometimes* actually inspired a single by the English band The Cure in 1981, which likely led to a small resurgence of reading the book around that time. The lyrics clearly do refer to the book, as part of the song goes, "By bedtime all the faces, the voices had blurred for Charlotte to one face, one voice" and the band refers to events in the book such as the 1918 Armistice. The book has also been produced on British television and (long ago) as an audiobook.

While Jessamy's cultural influence isn't so easy to trace on the large scale, readers have mixed reactions coming back to it after experiencing the book in childhood. A series of people on Goodreads have some recollections of reading the book as a child and then coming back to it when they were older. One person wrote, in a positive review, "I actually read this as a child when it first came out in the 1960s. All of the girls in my school fought over the lone copy our school library had." Another had a counterpoint: "Time-travel books featuring adolescent girls were all the rage in the mid-60s. Why, I will never know. As a girl in those days I gobbled them up; re-reading them today, they're all . . . of a mushness."[1] One thing to remember, though, is that in the 1960s there were few books that focused on children and especially on children who were not males. Even though the idea of time slips and girls in boarding schools has perhaps not aged well today, in the era these were groundbreaking looks at how girls experience the world and of their interpretation in it. And even today, some of the tropes still hold true—the friendships of girls don't change (even if the technology does), nor do the troubles of having to deal with meddlesome teachers and other adults. And there also is the struggle of growing up and trying to find one's way in an environment that can barely be controlled.

Today's children have much easier ways of finding friends and escaping reality, because (as long as their parents can provide for them) they have access to the Internet and mobile phones. This allows them to stay in touch with friends even if they are separated during the summer, although there are drawbacks, such as the risks of bullying, cybersex, and worse. I, like most children, would have appreciated the opportunity to "get out" a little more and to spend more time with friends, so probably on balance this technology is a good thing. But it very much depends on the child, and that's why these books still have so much resonance, since they do portray the two girls here as individuals who are struggling for their identity in a world where it is so difficult to find one.

1   Goodreads. (2019). "Jessamy, by Barbara Sleigh." Retrieved from https://www.goodreads .com/book/show/1127580.Jessamy

STAR TREK (1967-PRESENT)

Star Trek is one of those franchises that is so vast, so wide-reaching, that it is impossible to capture the extent of its drama and characters in only a chapter. But no matter what series you tune into, time travel is a uniting feature of the *Star Trek* universe. So let's dive in and look at a handful of episodes and try, as best as possible, to uncover the science.

Time Travel in *Star Trek*

Before we dive into time travel in *Star Trek*, understand that there are too many methodologies to cover in a short description. A quick search online shows competing versions of the fifteen or twenty best time travel episodes, so clearly there's a lot of material out there. And unfortunately, a lot of these episodes talk about time travel, but don't talk about how it happened. What we'll do here is round up some of the more prominent episodes and instances of time travel. If you'd like to see more, we invite you to sit and watch the several hundred hours of *Star Trek* shows available; you'll not only learn about time travel, but also a lot about philosophy, human relations, and Shakespeare. Yes, so much Shakespeare.

City on the Edge of Forever (The Original Series, 1967): While this is (as far as I can tell) the first use of time travel in *Star Trek*, it unfortunately goes a little deus ex machina on us—that is to say, the time travel device is something created by a powerful being and we're unable to see exactly how it is created, except by what some fans call "temporal waves." In any case, it's a cute episode. Doctor Leonard McCoy gets sick, he beams down to Earth, but gets caught in the wake of an ancient stone ring that can beam people to any time and place. He messes up the timeline, eliminates the orbiting

USS Enterprise, and forces his crewmates to go on a chase after him through Earth's past.

Star Trek IV: The Voyage Home (1986): The crew of the USS Enterprise responds to an alien distress call, this time from a ship that is unsuccessfully looking for whales—whales that are extinct on future Earth. While the ship wreaks havoc on Earth, the crew does this slingshot maneuver around the sun and somehow bends space and time far enough to go back through time—I suppose you could call it some sort of a wormhole, but I'm not sure that explains the strange, psychedelic images appearing on screen. At any rate, hilarity ensues when the characters find themselves in "present day" (1986) Los Angeles talking about "LDS" (that would be the hallucinogenic drug LSD) and asking what "exact change" means while riding a bus.

The Inner Light (Star Trek: The Next Generation, 1992): This bittersweet episode sees Captain Jean-Luc Picard hit by an energy beam from an alien machine in space. This transports him not only back in time, but also to live another being's life: Kamin. Picard spends forty years being Kamin and growing old among family and friends. Again, we don't know exactly how this time travel occurred, except for being told that the alien machine stores the memories of Kamin and his kind. It is notable that one can time travel in one's mind—after all, we all have memories—but to clearly experience the memories of another involves an element of mind-snatching or body-snatching on top of time travel. That's weird science.

Shattered (Star Trek Voyager, 2001): This is a creative way to recap all seven seasons of Voyager so far, if you were a fan new to the series and (before Netflix) you may have had few opportunities to catch up on episodes. The time travel in this case originates from a rift in space meeting with the ship's warp core—meaning, the engine that allows it to travel between distant stars at speeds faster than light. It's poorly explained how the ship gets out of sync with time, and the remedy (of course) involves the Borg—those creatures that are assimilations of many species across the cosmos. The science involves a lot of hand waving, but hey, the episode is a decent watch.

Magic to Make the Sanest Man Go Mad (Star Trek Discovery, 2017): Despite the dated use of "man" in the title, this is a cute look at how two crew members fall in love with each other over and over and over again when their ship gets caught in a time loop. The fun starts when Harry Mudd (Rainn Wilson) is accidentally beamed on board with, unfortunately for us science analysts, yet another time travel device whose functionality is not revealed. But we do learn that Mudd blows up the Discovery more than fifty times (and presumably, in more than fifty ways) in his quest to find the ship's fast-travel secrets: "There really are so many ways to blow up this ship—it's almost a design flaw!" he crows. For what it's worth, though, the episode does have hints of real-life science, including mentioning the importance of tardigrades—mini water creatures in real life that look a bit like hippos.

The Science of *Star Trek*

Star Trek is so influential a series that we often talk about technologies being inspired by *Star Trek*, or working like *Star Trek*, or appearing like *Star Trek*. It's hard in some cases to establish a causal link between invention in *Star Trek* and invention in real life. But for what it's worth, here are a few things that we see in older *Star Trek* shows that are real today.

Tablet computers: This began to become prevalent in *Star Trek: The Next Generation* (1987–1994), particularly among crew members of the Starship Enterprise. You would occasionally see a crew member sitting in their chair reviewing briefing notes on a flat computer, or strolling around the ship with information contained on this mini-computer. In the 1980s, the best most people had for flat computers was either a calculator or an Etch A Sketch (look it up). But in the last decade or so, most of us have come in contact with tablet computers—either devices wholly dedicated to reading, or more multimedia devices that can play video or surf websites.

Tricorders: Starting with the original *Star Trek* (1966–1969), medical doctors used little devices called tricorders to pick up basic medical information about patients. In fact, today the technology of diagnoses is leaping so quickly ahead that what I write here is likely going to be outdated very quickly. Artificial intelligence already assists doctors with reading radiation scans or sifting through databases. A real-life tricorder was commissioned after a contest in 2017 to seek a universal device to classify illnesses such as pneumonia, sleep apnea, or urinary tract condition.[1] What this really shows is the potential of humans and computers to work together to solve problems, with computers crunching the data and then humans taking the summary results to create the diagnosis.

Communicators: Well, that's basically every modern cell phone ever. We don't appreciate enough how much of a miracle it is to walk around with the Internet in our pocket, or on a wearable device. From most places in the world, it's possible to download the information you need via satellite, even when you're far from your house or even a large city.

Spaceships named Enterprise: The most famous Enterprise spaceship never actually went to space—it was a space shuttle replica that was designed to do landing tests, fully in the atmosphere. NASA wanted to name this spaceship "Constitution," but changed its mind after some determined *Star Trek* fans organized a letter-writing campaign. (Yes, it was the 1970s and yes, this does sound very dated today, but hey, it worked.) Enterprise did five landing tests, did some goodwill tours around the world, and today lives aboard a floating ship museum in New York City

1   Howell, Elizabeth. (2017, April 13). Live Long and Prosper: 'Star Trek'-Like 'Tricorder' Wins $2.6M Prize. Space.com. Retrieved from https://www.space.com/36461-star-trek -like-tricorder-medical-device.html

called the Intrepid Sea, Air & Space Museum. Another famous Enterprise ship came to a sadder end; Virgin Galactic's first test spaceship was named that, and the VSS Enterprise unfortunately crashed during a test flight, killing one pilot and seriously injuring another. Virgin did some crucial redesigns following the incident and has already performed several test flights in space[2] as of early 2019.

Also note that some *Star Trek* technology is still elusive. We haven't figured out human teleportation, although we know that quantum teleportation is possible if you're a teeny-tiny particle (*Honey I Shrunk the Kids*, anyone?) And warp drive is something that we're still dreaming of; if it ever is invented, the universe will open up in an incredible way and allow us to visit distant stars in days or months.

The Legacy of *Star Trek*

If you want to know how influential *Star Trek* is, just look at all of the people who became scientists, astronomers, astronauts, even doctors after watching the show as children. It's fun to see how real-life astronauts have interacted with *Star Trek* over the decades. Here are just a few examples: Leonard Nimoy (the actor who played Spock) died in 2015 while astronaut Terry Virts was in orbit aboard the International Space Station. In between other duties, Virts rushed to a window, displayed the "Live Long and Prosper" signal with his hands, and sent that out on Twitter. The post went viral and coincidentally, in the background was Nimoy's hometown of Boston.[3] Canadian astronaut Chris Hadfield got a

---

2   To be fair, it's "space" depending on the definition. These test flights have flown to an altitude of 55 miles, which is above the U.S. military's definition of space but still a little below the internationally defined Karman line at 62 miles.

3   Howell, Elizabeth. (2016, Sept. 7). Astronauts and 'Star Trek': How a TV Show Inspired Real Space Travelers. Space.com. Retrieved from https://www.space.com/33978-star-trek-inspired-real-astronauts.html

special tweet in space while starting his 2012–2013 mission. It was from Canadian actor William Shatner, who asked him in wonder if he was tweeting from space—a valid question since Twitter was a new technology at the time and astronauts only recently had half-decent Internet installed on the International Space Station. Hadfield's response was classic: "Yes, Standard Orbit, Captain. And we're detecting signs of life from the surface." For a day, it seemed, everyone on Twitter retweeted the conversation. Before long, it got the attention of numerous other *Star Trek* actors—Nimoy (then still living), George Takei, and Will Wheaton.[4] And there are other relations between *Star Trek* and astronauts as well. Nichelle Nichols participated in at least one astronaut recruitment campaign, which was a lovely gesture since she not only portrays an astronaut, but is also a person of color. And several astronauts have participated in *Star Trek* shows: Mae Jemison (the first African American woman in space), Virts (yes, he was on a show prior to tweeting about Nimoy in space) and Mike Fincke.

Modern-day audiences were lucky to see several *Star Trek* movies make a resurgence in the 2000s, although the pace appeared to fade quickly after the first one was released in 2009. A decade later, development on a fourth film permanently stalled. Fortunately, *Star Trek* continues on the small screen with *Star Trek Discovery*, an online-only CBS show that has received great reviews in its first and second season. Starring is Sonequa Martin-Green, who not only portrays a female person of color, but one who has a traditionally male name: Michael Burnham. Better yet, the show had its first openly gay couple and has been very strong in portraying diversity so far. In early 2020, during edits of this book, *Star Trek Picard* (a sort of *The Next Generation* sequel) made its debut—and a second season is already on the way.

---

4   Howell, Elizabeth. (2018, July 12). Star Trek: History & Effect on Space Technology. Space.com. Retrieved from https://www.space.com/31802-star-trek-space-tech.html

By the way, all of these mentions of *Star Trek* do not even begin to acknowledge the huge base of fan support that it holds. There are numerous fan productions in the works and already released, especially since the Internet allows anybody to host a show; for the most part, CBS and Paramount (who own the copyrights) have been allowing these productions, although one production resulted in a lawsuit that was settled out of court.[5] *Star Trek* actors at comic conventions continue to be large draws, not to mention that the *Star Trek* franchise has its own convention in Las Vegas every year. Fans consume *Star Trek* in every media imaginable: mobile games, comic books, video games, and (on occasion) old-fashioned books that you will find in bookstores or at the library. This is a lucrative franchise and truly has drawn fans from around the world, from all walks of life—and it appears to remain strong in Generation Z. Thank you, Roddenberry!

---

5 Howell, Elizabeth. (2017, Jan. 24). Lawsuit Over 'Star Trek' Fan Film Settled. Space.com. Retrieved from https://www.space.com/35436-star-trek-axanar-fan-film-lawsuit-settled.html

SLAUGHTERHOUSE-FIVE (1969)

If you like stories that completely break with narrative—the *Pulp Fictions* of the literary world—*Slaughterhouse-Five* is a novel that is fascinating on the first time and really rewards repeated readings. Its discussion of time travel may not be exactly what you expect, since there is no fancy equipment associated with it. But what it does show is how our lived experience shapes our perception, and that time travel can take place in unexpected ways.

The Science of Time Travel

Slaughterhouse-Five is a complex story, one that you may come across as required reading in high school, so this summary will only briefly touch upon the major themes. To really appreciate it, you're going to have to pick up the book. Billy Pilgrim, who is considered an "unreliable narrator" since his interpretations of what he sees are sometimes hard to square with, is working in the United States Army during World War II. It's an unfortunate place for him because he hates fighting, and is in fact trying to refuse to fight.

Pilgrim suffers a series of traumas in the war and eventually suffers what appears to be a form of post-traumatic stress disorder (more on that in a bit), where he has flashbacks and is described as "unstuck in time." In pure Vonnegut brilliance, even as Pilgrim resumes a normal life and recovers, he gets kidnapped by aliens known as Tralfamadorians. And here is where we start to get into the meat of time travel: As a Vonnegut fan site explains of these aliens, "They can see through time, and can live in any moment of time. They are fatalists, as they know how the universe will end, but cannot do anything to prevent it."[1] This is hard to unravel in a complex novel, so if I may, let's borrow from a more recent interpretation of living in all aspects of time at once: the 2016 movie *Arrival*. Here, aliens also come to Earth, but rather than engaging in an invasion, they instead engage in a lengthy conversation. One of the main characters comes to understand how these Heptapods experience time, which is not to see it as a long line with a beginning, middle and end. Instead, they can jump to any point on the line as they choose. We see this happening in the film as well. There's a key moment where a character is talking with somebody she has never met, using information from a future conversation of theirs to try to land a negotiation. That's because this character, just like Pilgrim, and the Tralfamadorians, has been unhooked from time.

Slaughterhouse-Five and *Arrival* are both discussing a kind of time travel that doesn't need machines or grand theories about how the universe works, but rather a kind of time travel that happens in our minds. Of course, if our minds are functioning normally, we can engage in time travel every day. Every time you remember or recall something, whether it be the location of your keys, the poem you need to memorize for English, or how you felt the first time you rode your bike, you are using different forms of memory. Memory can, simply put, be thought of as a kind of time travel. But we know from witness accounts that memory is also an unreliable form of time travel, as people find it hard to remember all the little details. That's why historical books are so fun to read—everyone disagrees on what happened.

We also can imagine Pilgrim and the Tralfamadorians being able to live in the future, because we all try to make predictions about the future ourselves. Here, of course, none of us are perfect

---

1   Kurt Vonnegut wiki. (n.d.) Tralfamadorians. Retrieved from https://vonnegut.fandom .com/wiki/Tralfamadorians

narrators. Think about all the mistakes made of history where we tried to predict the future, and failed. One fairly recent example is the financial crisis of 2008, which happened in part because of an unfounded belief that putting together risky mortgages in a package and selling it as an investment product would be a stable way of helping companies generate revenue. It did the opposite and several banks failed (or almost failed) as a result, dragging down the world economy along with them.

One could argue that everything above is not pure time travel—it's subjective travel, it's travel without machines, it's travel within oneself. Perhaps a greater lesson to be learned, though, is that as you travel into the future, you should be thinking about the lessons of history while doing so. Now, by this I don't mean that you have to be memorizing the Kings and Queens of England or studying the required battles in history class. What I mean is that you need to look for history that interests you. If you like skateboarding, for example, then seriously look into the history of it. Find out who pioneered your favorite moves, why it took so long for the Olympics to bring it in (it only started in 2020), and how different cultures around the world have embraced skateboarding, even in communities where skateboards are not easily accessible. Use skateboarding as a bridge to learn about different cultures, different peoples, advances in technology. And then just follow your interests from there. This isn't a fancy kind of time travel, but it's a time travel that you can use and enjoy for your everyday life, and hopefully, that makes it of use to you.

The Science of Post-Traumatic Stress Disorder

While many of us associate PTSD with war-weary veterans, in reality many of us can experience a form of it through our everyday lives. If you have ever been in a car accident, witnessed or experienced violence, or been in a difficult school or work environment, you could be subject to PTSD. The Mayo Clinic[2] explains that these symptoms typically happen a few weeks after the event, but sometimes can lie buried for years—they're talking about problems such as intrusive memories, negative changes in thinking and mood, and avoidance or changes in physical or emotional reactions. These can be difficult not only in your everyday tasks, but also in your relationships with others. The changes are so deep that they can even be seen in the brain. While there are several that can be seen, we'll just focus on a couple here. One example is the hippocampus, the part of the brain that is concerned with memory and remembering. Specifically, patients can experience reductions in the hippocampus volume, which leads to a few issues.

"PTSD patients with reduced hippocampal volume lose the ability to discriminate between past and present experiences, or correctly interpret environmental contexts," writes scientific and medical consultant Viatcheslav Wlassov on BrainBlogger.com. "The particular neural mechanisms involved trigger extreme stress responses when confronted with environmental situations that only remotely resemble something from their traumatic past." For example, he continues, "a war veteran cannot watch violent movies because they remind him of his trench days; his hippocampus cannot minimize the interference of past memories."[3] And here is where the flashbacks can come in, just like what Pilgrim was experiencing in *Slaughterhouse-Five*. But the term PTSD did not begin to be used until the 1970s, when veterans of the Vietnam War were complaining of its symptoms; the American Medical Association first recognized it in 1980.

---

2   The Mayo Clinic. Post-traumatic stress order (PTSD). Retrieved from https://www .mayoclinic.org/diseases-conditions/post-traumatic-stress-disorder/symptoms-causes /syc-20355967

3   Wlassov, Viatcheslav. (2015, Jan. 24). "How Does Post-Traumatic Stress Disorder Change the Brain?" BrainBlogger.com. Retrieved from https://www.brainblogger .com/2015/01/24/how-does-post-traumatic-stress-disorder-change-the-brain/

What's worse, the issues caused by PTSD can continue and worsen, in the sense that the person who is experiencing the trauma can go on to develop common mood disorders—think anxiety, or depression. Of note, this means that everyone's experience of PTSD is unique. It's a collection of symptoms individualized to your brain, based on your own lived experience of a particular event. So the approach for one PTSD patient may not necessarily translate to another. And since your brain is altered by what happened, just toughening up may not work. Which is to say, you may need more help.

If you're going through a tough experience and you need help, don't hesitate to reach out to a medical professional or to somebody at a free counseling service such as The Crisis Call Center at 1-800-273-8255.

The Legacy of *Slaughterhouse-Five*

Perhaps one of the most enduring legacies of *Slaughterhouse-Five* is that it is based on Vonnegut's own lived experiences during the war. He was a prisoner of war. He lived on hard work, poor rations, and occasional beatings. His time in captivity included being stuck inside of a meat locker, or a slaughterhouse. It's traumatizing even to read these words, isn't it? "When Kurt came home from the war, PTSD wasn't called by that name. If it went by any name, it was called 'shell-shock' or 'battle fatigue.' It often wasn't treated as a medical condition. Instead, it was considered a character defect or a failure of nerve," said John Krull, a friend of Vonnegut's, in *Indianapolis Monthly*.[4] "That meant veterans such as Kurt didn't get the help they might have needed to deal with a mental health condition that leaves survivors trapped reliving horrible events,"

---

4   Krull, John. (2019, April 1). The Enduring Legacy Of Slaughterhouse-Five. *Indianapolis Monthly*. Retrieved from https://www.indianapolismonthly.com/longform /slaughterhouse-five-50

he continued. "The consequences could be tragic. People with PTSD struggle more often with severe anxiety, with depression, with chemical dependency, and with other challenges. The US Department of Veterans Affairs reports that veterans commit suicide at more than twice the rate of the rest of the US adult population."

We also cannot uncouple the writing of *Slaughterhouse-Five* with the bombing of Dresden, which was intended to cripple communication and transport links within Germany (both for its military and its civilians). While the official count of people who died remains controversial, what is known for sure is that it still echoes as a problem of war, writes Jonathan Creasy in the *LA Review of Books*. "Billy Pilgrim becoming 'unstuck in time' makes the rather blunt point that massacres such as Dresden happened; they always have happened and they always will happen," he wrote. "The numbers, the facts and figures, need not be accurately represented, if the novel's purpose is, in effect, to prove that firebombing Dresden was not a unique act of brutality. It was one brutality in a long history of brutalities that seems to have no end in sight."[5]

We all carry scars in our lives, and it's hard for us to say that one lived experience is better or worse than another. Trauma, as I said earlier, is an immensely personal experience. For Pilgrim, perhaps one interpretation is that trauma led him to believe in life in the universe—that there were other beings whose powers far outstrip our own. This is the narrative that makes sense for him, and if we are truly to try to walk around in his shoes for a while, it's a good idea to sit with the narrative and to accept it—as unbelievable as it may sound to our own experience.

Part of the problem that people of color and minorities face is that people dismiss how they feel and what they are going through.

---

5   Creasy, Jonathan. (2015, March 6). "Revisiting 'Slaughterhouse-Five.' Retrieved from https://lareviewofbooks.org/article/revisiting-slaughterhouse-five/

A common complaint of people of color, for example, is that white folks claim that they look past skin color and don't really see race. That's too simplistic of a view, they argue—and I would hate to explain why here, as I am not a person of color and could never really, truly, understand what it's like to grow up in a society with so many historical and current barriers in place for them. But what I can do, what I hope we can all do, is to listen and to learn and to try to remember what they tell us and to carry it forward into our own lives. So when you hear somebody telling you something that sounds unbelievable, before you make a judgment, slow down. Really listen to them. Really try to imagine living through it. Perhaps, at the end, it's hard to "prove" their point of view, but even for a dementia patient, it is useful to understand what their reality may be. Because, let's face it, we all shape our own realities. That's what makes living in a society so interesting, and so difficult.

PART III
THE
BLOCKBUSTER
ERA

THE HITCHHIKER'S GUIDE TO THE GALAXY (1979–PRESENT)

We'll need to tread a little more lightly on the "science of time travel" in this series here, because *The Hitchhiker's Guide to the Galaxy* has morphed over the years and likely has some paradoxes of its own in the various retellings. That said, what is admirable about this sprawling series is just how many people it has touched in various media. It began as a BBC Radio 4 broadcast in 1978, moving on to novels, comic books, shows on stage, television, a video game (really!) and even a feature film, playing an important role in the science fiction culture of Britain.

The Science of Time Travel

One criticism of how time travel is used in the Hitchhiker's series is that it seems to be incidental to how the world-building is constructed. One writer in the *Guardian* pointed out that Hitchhiker's quickly began moving into various storylines in the radio and novel versions, which made it difficult for people to follow across the various versions.[1] Instead of just accepting that, Adams used the weaker plot device of parallel realities and time travel, the *Guardian* said:

Bent and bent back again, erased and re-recorded, the characters fade into a general crackle; there is something

---

1 Turner, Jerry. (2009, Oct. 3). "Does the Hitchhiker's Guide to the Galaxy still answer the ultimate question?" *The Guardian*. Retrieved from https://www.theguardian.com /books/2009/oct/03/hitchhikers-guide-galaxy-douglas-adams

Sisyphean about their movements, pushing and pushing toward resolution, only to be dropped right back to a new beginning, to push and push again. At the end of the final novel, *Mostly Harmless* (1992), the Vogons came back and destroyed Earth properly this time, killing off absolutely everyone; except in the later radio adaptation, in which there was instead a cheery reunion at the Restaurant at the End of the Universe.

Adams does distinguish himself from the fact by creating a satirical "grammar of time travel," which is mentioned briefly as Dan Streetmentioner's *Time Traveler's Handbook of 1001 Tense Formations*. As explained in *The Restaurant at the End of the Universe*:

> One of the major problems encountered in time travel is not that of becoming your own father or mother. There is no problem in becoming your own father or mother that a broad-minded and well-adjusted family can't cope with. There is no problem with changing the course of history—the course of history does not change because it all fits together like a jigsaw. All the important changes have happened before the things they were supposed to change and it all sorts itself out in the end.

This excerpt displays a confidence that the fabric of time resists any attempt by one person to alter it indistinguishably (which we'll see again in books such as *11/22/63*). It neatly gets around the ideas of paradoxes introduced in other time travel series, such as *Back to the Future*. But is this simply a way of establishing a firm framework to the universe, or is it a way that Adams used to try to get around incongruities in the various versions of the work? You'll have to decide that as you read it.

Improbably, though, the discussion of the manual then dissolves into a satire on how impossible it is to memorize the rules of English grammar, using the example of "Future Semiconditionally Modified Subinverted Plagal Past Subjunctive Intentional" as the point where most readers give up on reading it. The pages near the end of the book are actually left blank in recognition that people don't see any worth in going that far, which is a curious and cute statement for any author facing imposter syndrome. (Yes, I faced it down myself as I wrote this text.) This excerpt from the manual helps you understand the grammar complexities that you may face if you are living in one timeline and coming from another: "It [the manual] will tell you, for instance, how to describe something that was about to happen to you in the past before you avoided it by time-jumping forward two days in order to avoid it," Adams writes:

> The event will be described differently according to whether you are talking about it from the standpoint of your own natural time, from a time in the further future, or a time in the further past and is further complicated by the possibility of conducting conversations while you are actually traveling from one time to another with the intention of becoming your own mother or father.

Naturally, the statement is broad enough and vague enough to introduce paradoxes—what, you can become your own mother or father, and what the heck does that mean? But considering the hand-waviness of how other franchises deal with time travel paradoxes, you can't really criticize Adams for not only going that route, but also making fun of it. Unfortunately, his series gives few clues as to how time travel really "works," but it is refreshing that he includes a satire of it and isn't afraid to make fun of himself along with other franchises.

His approach to time travel reminds me a little bit of the 2009 movie *Star Trek*, which plays with the usual paradoxes and tropes of time travel. One of the most refreshing characters of the series is oddly, Spock Prime, the Vulcan alien who is living outside of time and trying to find a way to reconcile his life cycle with it. Instead of trying to get back in time, Spock Prime instead uses his wisdom to help others, which, really, should be a way that we all try to live. It's tempting to go back in time and try to fix our mistakes (or the things that people do to us), and in some cases that is a good thing to do. But where you can't, you can take from the experience and learn from it.

The Science of Randomness

There are many themes that come out of the Hitchhiker's series, but one of the more fun ones to explore is the feeling of randomness. Granted, his book does not come at it in a very scientific sense, but it portrays things happening for bureaucratic reasons (including the planned destruction of a planet near the very beginning, portrayed as just another consequence of an interplanetary construction project). Life seems to operate this way, and in fact randomness underlies much of our technology today.

So, how does randomness affect our everyday lives? One important point where it should be used is in matters such as population surveys. A survey is by definition a portion of a population, so it cannot be fully representative. While surveys should aim to be representative of certain groups within their samples, it's important to pick these representatives at random. For one thing, that creates a better degree of fairness, and for another that ensures that there are the best data points possible in coming to a conclusion. Surveys have profound impacts on everything from health care funding to political representation, so they must be designed properly.

It's also important to understand randomness when you are lucky enough to have money to invest, and you are hoping to put it in a vehicle such as the stock market. While most of us enjoy a measure of predictability, in investments it is also (usually) true that you need to take on more risk to enjoy the chance of a greater return. The stock market is tied to an overall market evaluation of multiple companies, but past returns are not an indicator of what will happen in the future. So we need to accept fluctuations—that there will be large dips such as what happened around 2008–2010, for example. But on the whole, investors say they usually get ahead by putting their money in a broad base of stocks in the stock market, accepting that randomness is the way to help their investments grow.

Cryptography, the science that protects our computing systems from attacks by hackers, also relies on randomness, because the quality of the numbers used often factor into how secure the system is. In fact, it is often by exploiting the *non*-random part of the system, the password, that hackers are able to break in. As much of a pain as it is to change your passwords regularly and to use sequences of letters and numbers that are hard to guess, it's an essential part of keeping your banking information and other life information safe from outsiders.

Randomness is also an important part of weather prediction (see our discussion of "the butterfly effect" earlier in the book on page 53). The premise to understand here is that you need to generate a range of events to best make predictions, with the idea that the "true" outcome will be one among the many dozens or hundreds or thousands that your system generates. This is, of course, not perfect, as sometimes the reality will fall outside of the predictions, but you increase the odds of making the right choice if you generate a number of predictions based on variable values.

The Legacy of *The Hitchhiker's Guide to the Galaxy*

While Hitchhiker's (at first glance) appears like a science fiction tale, a 2009 story in the *Guardian*, written on the franchise's thirtieth anniversary in the books, argues that it doesn't so easily fit in a category. Douglas Adams had experience not only writing for *Doctor Who*, but also for the famous British comedy *Monty Python's Flying Circus*, and the newspaper argues that it is more of the latter that shines through in Adams's writings. "Although the majority of the saga is set on distant planets, and contains plenty of real-life science and technology, at its core it retains a thoroughly surreal and very English sense of humor," the newspaper said.[2]

> The story's most fundamental themes[, it added,] include such reassuringly earthy—and earthly—subjects as cricket and towels. Yes, this is a story populated by such exotic characters as Gargravarr, the disembodied mind in charge of the Total Perspective Vortex, and Eccentrica Gallumbits, the triple-breasted [woman] of Eroticon Six. Yet there are also mentions of rich tea biscuits, Islington shops, *Which* magazine and American Express cards. Spaceships await deliveries of lemon-soaked paper napkins; aliens watch gameshows and take holidays in Bournemouth.

And lest you think this 2009 reference is somewhat dated, a writer in the *Irish Times* proclaimed the book "a hilarious map to life's terrain." The series, of course, heavily draws from preoccupations of the 1970s and 1980s, the newspaper says. "Being an unimportant dot in a petty system; the constant disappointment of 'progress' —and turning them into a layered, profound, imaginative, and hilarious story that transcended the era. Regardless of how many digital watch jokes snuck in."[3]

2   O'Dair, Marcus. (2009, Oct. 12). "The Hitchhiker's Guide to the Galaxy, 30 years on: why we should still be reading it." *The Guardian*. Retrieved from https://www.theguardian .com/books/booksblog/2009/oct/12/hitchhikers-guide-to-the-galaxy-douglas-adams
3   Hegarty, Shane. (2018, July 28). 'The Hitchhiker's Guide to the Galaxy': 40 years on. *The Irish Times*. Retrieved from https://www.irishtimes.com/culture/books/ the-hitchhiker-s-guide-to-the-galaxy-40-years-on-1.3575982

However, the representation and portrayal of females in this series is not great. It's a criticism that can be leveled at many of the same era, including *The Lord of the Rings* (which meant that the screenwriters for the massively popular Hollywood blockbuster essentially rewrote the role of Arwen for more representation). But focusing on Hitchhiker's, writer Jess Zimmerman pointed out that the first book had only one major female character in it, who was not shown in the most flattering light.[4] "It's not that Douglas Adams didn't try with Trillian," she writes. "He gave her two higher degrees, in math and astrophysics. He made her the nominal romantic partner of one of the characters, yes, but not the sympathetic Everyman hero; Arthur has designs on Tricia when they first meet, but he never 'gets' or even pursues her, at least not while Adams was alive." The worries come more with Trillian's role in the series as a whole:

> Trillian doesn't get a lot of laugh lines, or a lot of interior landscape; most of the time, she's seen from the outside, filtered through a male observer. Until her alt-Earth version appears in book five, there's really only one point where the omniscient narrator dips into her psyche and gives us a sense of Trillian as a subject—and it's when she's deciding to leave Zaphod. And it's that lack of subjectivity—the fact that even attempts to make Trillian a more dimensional female character still stop short of making her a fully-realized person—that leaves her, and the series, vulnerable.

While Adams was by far a cultural product of his era, there are parts of the humor of Hitchhiker's that remains timeless. It's still a

---

4   Zimmerman, Jess. (2015). The Problem With Trillian: 'Hitchhiker's Guide' and Me. The Toast. Retrieved from http://the-toast.net/2015/09/01/the-problem-with -trillian-hitchhikers-guide-and-me/

common joke to reference his proclamation that the "answer to life, the universe and everything" was the number forty-two, which is probably as good an explanation as any out there. (And which leads to an endless number of jokes when somebody you know turns forty-two years old.) And there is something endearing about the advice to not panic *and* not forget your towel, which symbolizes both flexibility in life circumstances and the need to be prepared. So despite some of his characters not aging well, it's likely much of the humor will be relevant for a long time to come.

TERMINATOR (1984-PRESENT)

Honestly, it's hard to think of a better blockbuster franchise than Terminator. Its tale about cyborg killers, robot rights, and time travel seem more prescient today (nearly forty years later) about what to expect in the future. What is the place of robots in society? Will allowing them to control our weapons and equipment lead to a global war, or are they better able to handle themselves than humans?

The Science of Time Travel

As fans well know, *Terminator* has spawned multiple sequels, with the latest (*Terminator: Dark Fate*) that premiered in 2019. We won't recap the whole franchise here in the plot summary, although we will pull time travel elements as they are explained in the films. *Terminator*, which multiple sources say was influenced by an episode called "Soldier" from *The Outer Limits*, written by Harlan Ellison and adapted from an Ellison short story "Soldier From Tomorrow," follows a time-traveling robot.[1] The first movie opens in 1984 Los Angeles—the present day when the film was released—and shows the Terminator (Arnold Schwarzenegger) on the hunt for women named Sarah Connor, as listed in the telephone directory. Shortly after the Terminator's arrival, a soldier called Kyle Reese (Michael Biehn) arrives and also goes on the hunt for a particular woman who has the name Sarah Connor (Linda Hamilton). Kyle eventually meets up with Sarah and explains that

---

1   Cronin, Brian. (2018, June 30). Movie Legends: How Harlan Ellison Got an Acknowledgement on Terminator. CBR.com. Retrieved from https://www.cbr.com /terminator-harlan-ellison-credit/

in the near future, an artificial intelligence defense network called Skynet will be "self-aware" and decide the best course of action is nuclear war, because after all, this was the 1980s and nuclear discussions were very much a thing between Russia and the United States as a part of the Cold War. The Terminator wants to kill Sarah because her unborn son, John, will be the one leading the human resistance in the future.

While the first movie is pretty much the Terminator chasing Sarah, and Kyle trying to protect her, future movies deal more with robot rights. A particular favorite of fans is *Terminator 2: Judgment Day* (1991), for example, which shows a kinder version of the Terminator going back in time to protect both Sarah and John (Edward Furlong). Connor, obviously, has trust issues given another of the Terminator's line was so dedicated to trying to kill her. But John argues the Terminator should be given a chance, talking his mother out of smashing an important part of the computer's memory and teaching the Terminator human expressions and the fun of high-fives.

How exactly do the Terminator and his counterparts travel back in time? According to the Terminator wiki fan page, the secret sauce lies in something called Time Displacement Equipment (TDE), also known as a Time Field Generator.[2] Located in an underground complex in Los Angeles, it doesn't look like many other time machines we're used to—it's actually a huge room about the size of your gym in high school. But instead of being a venue to play basketball, it's a location with a hole in the center of the floor. Inside that hole are three large chrome rings hovering in a magnetic field. "Individuals traveling through time must first be covered with a conductive substance so that the time-field will follow their outline, then step into the hole at the center of the inner

---

2   Terminator Wiki. (n.d.) "Time Displacement Equipment." Retrieved from https://terminator.fandom.com/wiki/Time_Displacement_Equipment

ring, where they will be suspended like the rings," the wiki says. "As the TDE is started up, the rings rotate around each other on different axes like a complex gyroscope, and the floor splits open like wedges of a pie which pull back from the center." Eventually, the rings spin super-fast to create an electrical charge. When the charge is high enough, a flash of light is visible and the traveler gets sent through time, either forward or backward. Unlike other time travel machines, this TDE is only a one-way trip. Certainly, though, if you travel to a location in time where another TDE is available, you can make your way back to where you came from.

Let's first start with some of the time-travel paradoxes brought up in the *Terminator* series, and with apologies, you are going to see some mild spoilers in here so we can properly go through them. A writer for *Vice* spoke with a theoretical physicist about some of the causal-loop problems, such as:

- John Connor's dad is actually Kyle, who traveled to the past to protect Sarah. And it was future John Connor who ordered Kyle to initiate the trip.
- The future was supposed to be protected when John Connor was born (in between Terminators 1 and 2), but in Terminator 2 writers changed the scenario and declared that the future actually wasn't safe after all as there were still spare parts of future Skynet available in present-day 1991.[3]

Vice then points to a theory in quantum mechanics—the science of the very small—known as the "many worlds interpretation." This theory holds that while we are living in our own world and time, there are other parallel worlds existing simultaneously where

---

3   Pearl, Mike. (2015, July 1). We Asked a Theoretical Physicist How Time Travel in the Terminator Movies Works. *Vice*. Retrieved from https://www.vice.com/en_ca/article/dp5pxv/we-asked-a-physicist-how-time-travel-in-the-iterminatori-movies-works-721

different things happen. For example, in the reality where you read this book, you know enough English to understand the English translation. But the "many worlds" could theoretically say there are different versions of yourself in different worlds that could understand any language that humans speak, from Russian to Somali.

With the caveat that this author is not a theoretical physicist, it seems that the "many worlds" idea as it relates to time travel is, to say the least, controversial. David Deutsch assumed a scenario where a quantum system in state S1 interacted with its older self, according to the *Stanford Encyclopedia of Philosophy*.[4] The quantum system is now in a new state (called S2) that travels back in time (becoming S3). S3 then interacts with the younger self and becomes S4. But there are issues with this idea, argue article authors Frank Arntzenius and Tim Maudlin. Once two states interact, they argue, it will create an "entangled state." This means that it's very difficult to predict what happens after the interactions. One of the logical follow-up questions, the authors add, is "Will the state of the younger system remain entangled with the state of the older system as the younger system time travels and the older system moves on into the future?"[5] It's a tangled thread of scenarios which also seem difficult to reconcile with time travel, but don't let that interfere with your fun watching the *Terminator* series (or any other time-travel series, for that matter). Hey, I got a lot of enjoyment out of the movie *Gravity* even though there was no possible way the astronaut could have traveled between spaceships in different orbits—it would have used too much fuel. So go ahead and make fun of the science in *Terminator*, but appreciate it for its other merits: the plot, the characters, and the amazing special effects that still stand up, even for the 1984 film.

---

4   Arntzenius, Frank and Maudlin, Tim. "Time Travel and Modern Physics". (Winter 2013). *The Stanford Encyclopedia of Philosophy*. Edward N. Zalta. Retrieved from https://plato .stanford.edu/archives/win2013/entries/time-travel-phys/.
5   *Ibid.*

The Science of Algorithms

Let's get back to that idea of Skynet creating a nuclear holocaust. It's argued that the system did this out of some sort of malice (see the self-awareness argument). We're not here to argue if computers can or cannot become self-aware, but we can point to computers making questionable decisions because they are constructed to follow algorithms—rules or processes put in place by humans (or which the computer learns itself based on an initial set of instructions created by humans). There are all sorts of examples of computers making morally ambiguous "decisions" through algorithms, although we must make it clear the computer isn't doing this deliberately—it's due to a mistake by humans or because the humans inserted something. One example that really shook me was YouTube's proprietary algorithm, which is built on trying to get humans to click, to engage, to react. One former YouTube engineer, Guillaume Chaslot, told the *Guardian* that the recommendation algorithm "is not optimizing for what is truthful, or balanced, or healthy for democracy." While it's tough to prove or disprove Chaslot's words—when the article was published in 2018, the algorithm was not released to the public for obvious business reasons—the *Guardian* points to problematic examples such as toddlers shown adult-only content while using what is supposed to be a kid-friendly version of YouTube.[6]

So, how do computers make choices through algorithms? How Stuff Works has a clever example where you're trying to pick up a friend from the airport. You could either suggest a taxi, ask the person to give you a call ("the call-me algorithm"), suggest they rent a car, or (if you're feeling particularly angry on a cold winter day, perhaps) tell them to take the bus. Each algorithm will get the

---

6   Lewis, Paul. (2018, Feb. 2). 'Fiction is outperforming reality': how YouTube's algorithm distorts truth. *The Guardian*. Retrieved from https://www.theguardian.com/technology/2018/feb/02/how-youtubes-algorithm-distorts-truth

friend to your house, but there are obviously different travel times and costs involved. The bus is the slowest and cheapest method, while the taxi is fastest and most expensive.[7] Humans make choices based on a variety of factors involving time, money and energy, and as John Green keeps reminding viewers in his Crash Course: Navigating Digital Information YouTube series,[8] humans also make choices when programming computers to make decisions. If you're a programmer, ask yourself: What is your bias as you create your creation? Is it to help someone make money, is it to perform a public service, is it to educate, or what is it? Because the end goal you have in mind is ultimately what the computer will perform. This means in today's world, ethics in computer programming has as much importance as engineers designing bridges, or journalists writing articles for public consumption.

If you're unsure how to proceed, often your workplace will have an ethics code. Or you can consult the more well-known ones, such as the Association for Computing Machinery's Code of Ethics and Professional Conduct.[9] I particularly like the first stipulation at the top: "Contribute to society and to human well-being, acknowledging that all people are stakeholders in computing." To be honest, that's a lesson that all of us could use in life, to use our good to benefit others. What was it some of the old Spiderman movies said? "With great power comes great responsibility."

The Legacy of *Terminator*

There's no greater thing to say about the legacy of *Terminator* than to give the good news, in that another of the series came

7   How Stuff Works. What is a computer algorithm? (n.d.) Retrieved from https://computer.howstuffworks.com/what-is-a-computer-algorithm.htm

8   Green, John et al. (2019). Crash Course: Navigating Digital Information. YouTube. Retrieved from https://www.youtube.com/watch?v=L4aNmdL3Hr0&list=PL8dPuuaLjXtN07XYqqWSKpPrtNDiCHTzU

9   Association for Computing Machinery. (2018). ACM Code of Ethics and Professional Conduct. Retrieved from https://www.acm.org/code-of-ethics

out in 2019. Called *Terminator: Dark Fate*, it happily starred Arnold Schwarzenegger and Linda Hamilton and even included the involvement of franchise creator James Cameron for the first time since *Terminator 2: Judgment Day* in 1991. Having all of these original players is a positive sign, since it means there is some commitment to the franchise that may even go beyond *Dark Fate*. As long as the movie did its job in appealing to a new audience, we may see more *Terminators* coming in the 2020s. After all, the original movie was set in 2029, so maybe we'll be lucky enough to see a new *Terminator* coming out that year. (And hopefully, Skynet won't have killed us by then.)

In the broader sense, the *Terminator* franchise launched the careers of both Schwarzenegger and Cameron. If you enjoyed *Alien*, *Avatar*, *Titanic*, or any number of sci-fi themed films, you can thank *Terminator* for giving Cameron the cash and the Hollywood stature to get these things financed and into your local cinema.

But there's still something about these older *Terminator* movies that make you want to come back again and again. There are the timeless lines: "I'll be back" is much parodied, of course, but it still sounds fresh with Schwarzenegger's performance. There are the special effects: *Terminator 2*'s melting and reforming of metal still awes even thirty years on, for example. And let's not forget that Schwarzenegger, despite being a young actor working in a foreign language on these older films, has a humanity that shines through the accent and the stiff role he was given. Who didn't cry when the Terminator told Connor, "Come with me if you want to live."? This gives the movies high playability and replay-ability value even decades hence—meaning they will likely remain popular for the foreseeable future.

Any newer films will have to come close to the mark of the originals to remain current. There are so many more action films and also, so many more countries in the movie market than in the 1980s and 1990s. Schwarzenegger and his costars need to play well not only in the United States, but in the now-powerful Asian markets of Japan and China and India. It might be that the franchise needs to hand off the actors to somebody of a more diverse background to keep the series going and to look fresh in the eyes of younger viewers. Could we have more female Terminators, or somebody of a minority background? That's really up to the franchise creators. But it might be something to consider as Schwarzenegger ages and the audience becomes less and less familiar with the original *Terminator* creations.

BACK TO THE FUTURE (1985-1990)

Back to the Future is such a famous series that it almost becomes its own inside joke. The silver DeLorean, the ever-present expressions "Great Scott" and "Heavy," and the wild look in Doc Brown's eyes were some of what kept fans coming back again and again. What also makes the series outstanding is that every individual actor is a great comedian. But how about its time travel—does it still stand up?

The Science of Time Travel

While there are three movies, we're going to focus most heavily on the first and most famous of the three: 1985's *Back to the Future*. Marty McFly (Michael J. Fox) is a teenager living with typically clueless parents. His biggest problem is trying to get a reliable car to take his girlfriend out on a date. He's late to school and prefers playing guitar to studying, prompting the principal to call him a "slacker." You know, typical high school stuff. But it's his friendship with Doc Emmett Brown (Christopher Lloyd) that propels Marty into a new trajectory, even into another generation. Brown has modified a DeLorean car into a time machine, using a device called a flux capacitor to make traveling in time possible. After a tragic incident in a parking lot, Marty finds himself in the year 1955. He accidentally gets in the way of his parents meeting, and must make them fall in love before his existence is wiped away forever.

Let's start with the time machine. What many viewers may not know is the DeLorean was a real-life luxury car in the 1980s, designed by Italian Giorgetto Giugiaro and manufactured by the DeLorean Motor Company. (There's another company today that supports DeLoreans, but it's not this original one—it went bankrupt long ago.) You can in fact still buy a DeLorean today, and there are a few that make the circuit at comic-cons, complete with the iconic display of times and dates on the dashboard time machine panel. Oddly enough, the iconic time machine and car wasn't originally in the script. "It took us a while to work out how Marty McFly would travel through time," said cowriter and producer Bob Gale in a 2014 interview. "We knew it had to be an accident. One of the conventions of time-travel movies is that it can't turn out well if the character is doing it for any kind of personal gain. But there was no DeLorean originally. In the first two drafts it was a time-chamber apparatus."[1]

The DeLorean traveled back in time using a device called a flux capacitor, which rode just behind the passengers. It constantly breaks down in the films due to fuel issues (requiring creative ways to get it jump-started again, but we won't spoil how). As the first film explains it, Brown first thought of the idea after (stupidly) standing on his toilet to hang a clock, on November 5, 1955. He slipped, bumped his head, and was knocked out. After coming to, presumably with a nasty concussion, he drew out the design and spent the next thirty years trying to make it real. How it goes back through time is . . . complicated. The Fandom wiki page for the *Back to the Future* franchise explains that it includes temporal demodulation coils that open up a rift in time.[2] Every fan of the movie knows what that requires: a *lot* of speed and power. The DeLorean needs to travel at 88 mph while getting a jolt of 1.21 gigawatts of power, which originally required a (very illegal) supply of plutonium to accomplish.

1   Gilbey, Ryan. (2014, Aug. 25). Interview: How we made Back to the Future. *The Guardian*. Retrieved from https://www.theguardian.com/film/2014/aug/25/back-to-the-future-michael-j-fox-christopher-lloyd-how-we-made

2   Futurepedia, the BTTF Wiki. (n.d.) Flux capacitor. Retrieved from https://backtothefuture.fandom.com/wiki/Flux_capacitor

And there, our knowledge of flux capacitor comes to an end. The films never really explain how the device allows you to travel through time, so it's hard to point to any scientific analogy to find out how true (or false) this time-travel method is. But while we're a little confused by this time travel method, we can tell you that there are real-life flux capacitors available. They're just not made for time travel. In 2018, Engadget reported on a "new type of electronic circulator, which can control the directional movement of microwave signals."[3] This device actually has a couple of designs, although one of those designs gets its inspiration from the flux capacitor. The physics is complicated, but essentially this device takes advantage of quantum mechanics (the science of how fundamental particles behave) to force microwave signals to move around a circuit in only one direction. Unfortunately, time travel won't be a result of this device, but the researchers (who published their work in *Physical Review Letters*[4]) foresee some applications in quantum computing (which would allow calculations to happen much faster), as well as improving the signals used by radar, Wi-Fi, and mobile.

So who knows, maybe one day you'll be using a flux capacitor daily to get better connections to the Internet . . . and funnily enough, it foreshadows a cute skit that Fox and Lloyd performed on a late-night show in 2015. On the day when they're supposed to travel "back to the future," they appeared on stage on Jimmy Kimmel to learn what 2015 is really like. Upon hearing that there are no flying cars and that cronuts are one of the top achievements of our era, Lloyd remarks, "The technological and cultural

---

3   Holt, Kris. (2018, May 28). Scientists invented a real-life flux capacitor, but not for time travel. Engadget. Retrieved from https://www.engadget.com/2018/05/28/flux -capacitor-back-to-the-future-quantum-computing/

4   Müller, Clemens et al. (2018, May 25). Passive On-Chip Superconducting Circulator Using a Ring of Tunnel Junctions. *Physical Review Letters* 120(21). doi: https://doi.org /10.1103/PhysRevLett.120.213602

achievements of this era are somewhat underwhelming." But they do seem intrigued by smartphones.[5]

The Science of Self-Confidence

While the time-travel science is a little lacking, we can go a little bit into another undercurrent of the original *Back to the Future*: how to get a little more confidence in yourself. Marty isn't convinced that he should send his guitar-playing to a record label, because what if he's rejected? And upon going back to 1955, he discovers that his dad struggles with many of the same self-confidence issues. Is it worth going for the girl that he has a crush on, even though she may not feel the same way? Should he dare to share the fictional short stories he's working on, even though the publishing business is tough? The simplest way of addressing this is first of all, to understand your thoughts. Our memories allow us a way of time traveling through our mind. This means that when we make a decision to do something (or not to do something), we're engaging in a bit of forecasting—in deciding whether or not we have the ability to do something. But often, this "ability" is born purely from belief, as *Forbes* puts it: "Confidence is not a fixed attribute; it's the outcome of the thoughts we think and the actions we take. No more; no less."[6]

How can we convince ourselves we are more likely to succeed than fail? Perhaps the most important thing is to cultivate what experts call a "growth mindset." There's been a lot of research in this field from Stanford University's Carol Dweck,[7] among many other scientists, backed up by research in past decades showing

5   Jimmy Kimmel Live (Facebook). (n.d.) Marty McFly & Doc Brown Visit Jimmy Kimmel Live. Retrieved from https://www.facebook.com/watch/?v=10156689301233374

6   Warrell, Margie. (2015, Feb. 26). Use It Or Lose It: The Science Behind Self-Confidence. *Forbes*. Retrieved from https://www.forbes.com/sites/margiewarrell/2015/02/26/build-self-confidence-5strategies/#72b82746ade9

7   Dweck, Carol. (2006). *Mindset: The New Psychology of Success*. New York: Random House.

that the mind can remain flexible (it has "plasticity," as the experts say) to allow for learning from experience. That plasticity persists even as you age, which means that yes, if you do want to go back to school in your forties (or even in your eighties!) to finish your university degree, by all means do go ahead.

We can't guarantee that changing your mindset will make you a successful rock star (which Marty wanted) or fiction author (as his dad wanted), but with some creativity, in many cases you can build a life that includes elements of your passion. If you love music, you could move into related fields such as being a DJ for corporate events, or perhaps, get a "day job" that pays the bills and do the music in your spare time on the weekends and evenings. It all depends on what kind of life you want to build, within the finances and other limitations you must work with. What are you going to work on next?

The Legacy of *Back to the Future*

What is the legacy of this iconic movie? One thing that strikes audiences upon watching it today is overall, it is fresh, but there are some issues with how some of the more sensitive items are handled. Near the end of the original film, something negative happens to one of the main characters and her assaulter barely gets punished. In this more sensitive era of #MeToo (which rocked Hollywood to its core and led to the end of several people's careers), it's clear that this is a more aged way of dealing with problems—to sweep them under the rug. Modern audiences, especially modern preteens and teens, may find it a bit difficult to accept how things are handled. On the more positive side, perhaps this helps us realize how far we've come since the era portrayed in the film (the 1960s) or the era of the film (the 1980s).

The most unexpected legacy of this franchise, stemming from how it lifted the career of Michael J. Fox for a long while, is awareness for Parkinson's Disease. Fox and several of his *Family*

Ties costars ended up with this disease, whose origins and cure still elude us. Fox continues to act and he is also a tireless advocate for charity. If you've had the chance to sit in a DeLorean at a comic con, for example, more than likely the funds raised went to Parkinson's Disease. Fox has testified before Congress, written books, and done a lot of media interviews about the disease, and people worldwide thank him for his work.

But there's still much to enjoy about the film on its own merits. The witty dialogue still sticks; although "Heavy" and "Great Scott!" sound dated, in a way it gives the trilogy its own language and it instantly marks you as a fan if you use these phrases. All of the actors are excellent comedians in their own rights, the technology portrayed in the film does not seem too far off from that of modern day (except, sadly, for the flying cars), and the soundtrack still is a fun ride after all these years.

Another lovely thing about the film is how it shows the potential for what a "slacker" can be in the world. It shows that high school is not the full answer to success in life, or even as a teenager. It shows that tinkering in a garage or a dorm room can lead to great adventures, something that definitely inspired the careers of the founders of Apple and Microsoft and Facebook, among the brands that influence us today. And most of all, the film shows us that a good life is a life in which there is somebody you can count on. I mean those folks that really get you, even if you don't fit into societal expectations and even if you're not on the path that most high-schoolers or university folks are on. We're all going to get to the Great Beyond at the end of our lives, so why not dare to be a little experimental? As long as you're happy, and as long as you're productive enough to take care of yourself and to give back to society in some way, I assure you, you're going to be fine.

TIMELINE
(1999)

Michael Crichton is the popularizer of many a science fiction franchise. Best known for the masterpieces of *Jurassic Park* and *ER*, he created several viable storylines that are heavily based in "hard science," that is, science that can be tested and found in the real world, even if the implications are fictional in Crichton's universes. Here, we'll focus on *Timeline*, which focuses on the world of time travel.

The Science of Time Travel

I'm a sucker for a good medieval story (even those that stretch reality a little too far), so I really enjoyed the mystery surrounding a French village in *Timeline*, which archaeologists (mostly graduate students who seem more focused on, well, having a good time than doing work) are examining. They're doing their usual (somewhat half-hearted) cataloging of artifacts when they discover an artifact that improbably seems to be including modern-day materials. After a few more wrong turns, the young team of archaeologists discover that there is a huge company trying to make use of quantum technology and a form of digital human cloning. It turns out that you can reconstitute people at distant points in the past, including the scene of the European Hundred Years' War. However, this situation creates genetic glitches and a team of grad students need to solve it while surviving the brutal world of 1300s France.

Crichton's genius here is anticipating the use of quantum technologies in the year 1999, which is very prescient of its use today in items such as cryptography. Here, however, he was focusing on then recent advancements showing that photons—essentially, particles of light—could be transported *Star Trek* style between laboratories.[1] Unfortunately, photons can do a lot of stuff that we can't do (at least as far as our physics can show us) because they (practically speaking) have no mass. This allows them to, well, travel at the speed of light and also engage in teleportation, something that we as people have not figured out how to do in either space *or* time. It's also interesting that he focuses on genetic alterations as a side effect of the time-travel process. While we can't speak to how real that may be in time travel, it is a very real concern for people who are exposed to radiation for long periods of time. One implication is that in space travel, astronauts are subject to lifetime limits in space because there, they are less shielded from radiation than we are here on Earth. In fact, some astronauts have argued that those limits are discriminatory, as the calculations (at least how they were portrayed in the mid-2010s) favor biological (not gendered) males over females.[2]

While we're still struggling with teleporting items besides photons, for what it's worth, our advances in this field have gone far beyond the laboratory. In 2016, two teams managed to move quantum teleportation outside the laboratory. And in 2017, in true *Star Trek* style, a Chinese team transported a photon from the Earth to a satellite (yes, a satellite) that were about 310 miles apart. This is made possible by using a principle known as quantum entanglement, which is a set of photons or other quantum objects that form simultaneously at a point in space. "In this way, they share the same existence. This shared existence continues even when the photons are separated—meaning a measurement on one immediately influences the state of the other, regardless of the

1   McCullagh, Declan. (1999, Sept. 12). "Crichton's Bad Timing." *Wired*. Retrieved from https://www.wired.com/1999/12/crichtons-bad-timing/
2   Kramer, Miriam. (2013, Aug. 27). "Female Astronauts Face Discrimination from Space Radiation Concerns, Astronauts Say." Space.com. Retrieved from https://www.space.com/22252-women-astronauts-radiation-risk.html

distance between them," *Futurism* said in a 2017 article.[3] "This link can be used to transmit quantum information by 'downloading' the information associated with one photon over an entangled link to another photon. This second photon takes on the identity of the first. Voilà. Teleportation."

We've only known of one set of human twins who were separated between space and Earth, which is the Kelly brothers—particularly during the one-year mission of Scott Kelly in 2015–2016 while his retired-astronaut brother Mark remained at home. The twins experienced subtle changes in gene expression between them (not changes in DNA itself, but gene expression, and Scott showed other signs of temporary stress in his immune system, bone activity, and other factors.[4] But clearly no body-switching or changes of identity took place, so we have a ways to go.

While the uses of teleportation for human bodies is suspect, we can use this same principle for digital cryptography. As *Popular Science* explains, "I start by encoding it in binary using the states of a group of photons (I could say clockwise is 1, counter is 0). I can securely share the digits thanks to a trick that occurs when two particles of the same type interact: entanglement. If Photon 1 spins clockwise, Photon 2 whirls counter. If one changes, so does the other—no matter how far apart they are." The magazine continues:

I can break up such a pair. One stays with me, one beams to you in a ray of light—and know their states will always be complementary. This means you can infer the info stored

3   Marquart, Sarah. (2017, July 10). "Scientists Just Teleported a Photon from Earth to Orbit for the First Time." *Futurism*. Retrieved from https://futurism.com/scientists -just-teleported-a-photon-from-earth-to-orbit-for-the-first-time

4   Lewin, Sarah. (2019, April 11). "Landmark NASA Twins Study Reveals Space Travel's Effects on the Human Body." Space.com. Retrieved from https://www.space.com/nasa-twins -study-kelly-astronauts-results.html

on my Earthly particle by measuring your own. All I have to do is wait for my half of the couple to take on the same state as a third photon—one that I encoded with a digit of binary—and tell you to examine your own. In an instant, it's turned into a precious passkey.[5]

While we can't use the quantum teleportation to transport ourselves, luckily, we can use it to transport information. And in a world where it's getting harder and harder to evade the hackers, we hope that this technology will make things just a little bit easier.

The Science of Trebuchets

The movie version of *Timeline* (which is surprisingly compelling as well as campy) shows a lot of trebuchet-firing in a climactic moment of the film, which means that a little look at these catapults seems apt even in the middle of a discussion about time travel. While trebuchets are portrayed as a French technology, their origins (like so many other European works!) came from China. They appear to have been used as far back as the third or fourth century BC. Their first known use in the west was by the Avars and the Byzantine Empire, many centuries later. (The word "trebuchet" is French, however, although we have many other names for this technology.)

Trebuchets can send their payloads an astonishing distance away, and can send aloft an amazing amount of mass. This came in handy during the medieval era, when they were used to transform walls to rubble or at the least, to get over the high ramparts and into the armies defending them behind the barrier. "A trebuchet works by using the energy of a falling (and hinged) counterweight

5   Francis, Matthew R. (2019, March 29). "Quantum teleportation is real, but it's not what you think." *Popular Science*. Retrieved from https://www.popsci.com/quantum-teleportation-real/

to launch a projectile (the payload), using mechanical advantage to achieve a high launch speed," explains Real World Physics Problems. "For maximum launch speed, the counterweight must be much heavier than the payload, since this means that it will 'fall' quickly."[6] While the physics of a trebuchet swing are a little beyond the scope of this book, Real World Physics explains that one way of thinking about it is an upside-down golf swing. In both cases, it's understandable that you are trying to launch something for a long distance.

To really have some fun learning about trebuchets, I highly recommend the slightly dated (but eminently astounding) *Nova* episode "Secrets of Lost Empires: Medieval Siege." Seeing modern engineers and scientists wrestle with millennial-old ideas is always a fun concept, and seeing how far these trebuchets can launch stuff is a both terrifying and awesome sight.

The Legacy of Crichton

I had a brief personal connection with Crichton's work when I interviewed one of the screenwriters of *Jurassic Park*, David Koepp, after Koepp wrote a book called *Cold Storage* about an Earth microbe that takes on strong properties despite having a stay in space. Koepp said he was influenced by Crichton's work, and his book is another take on the resilience of life (which fascinated both himself and Crichton). In fact, Koepp said he enjoys looking at this in real life, including when he was wrestling to remove a persistent weed from his backyard. "In an interview, he told Space. com about wrestling with an Asiatic vine that blew into the New York City region on the back of Superstorm Sandy in 2012," I wrote. "The vine can grow 6 inches (15 centimeters) in a day, he said with

---

6   Real World Physics Problems. (n.d.) Trebuchet Physics. Retrieved from https://www
    .real-world-physics-problems.com/trebuchet-physics.html

wonder, adding that he's been pulling it up repeatedly by the roots only to watch it revive again and again."[7]

While Crichton's influence on pop culture today is not as immediate as before his death, *Vanity Fair* pointed out that there was a period in his lifetime when he reigned over many trends of the moment.[8]

"There has never been anyone quite like him in the history of the movies," the magazine wrote.

> In his lifetime Michael Crichton wrote eighteen major novels, most of them bestsellers, including *The Andromeda Strain*, *The Great Train Robbery*, *Jurassic Park*, *Congo*, *Disclosure*, and *Sphere*. His books have sold more than two-hundred million copies worldwide, and thirteen of his novels were made into major films, many of them huge financial successes (the *Jurassic Park* juggernaut alone has earned more than $3.5 billion worldwide). He also directed seven films (including *Westworld*, *Coma*, *The Great Train Robbery*)—all of this making Crichton rich beyond the fantasies of most writers.

Crichton, however, found that he could not write dozens of novels without controversy—and indeed, in some sense he seemed to enjoy the challenge of science and the models. *Vanity Fair* focuses on at least three novels that dealt with tough topics: *Disclosure* (feminism and sexual harassment), *Rising Sun* (Japanese corporate domination of technology), and *State of Fear* (global warming). There's much to be said of any of these topics, and some of their

---

7   Howell, Elizabeth. (2019, Sept. 5). "How Skylab's Demise Inspired 'Jurassic Park' Screenwriter's 1st Novel 'Cold Storage'". Space.com. Retrieved from https://www.space.com/david-koepp-interview-cold-storage-book.html

8   Kashner, Sam. (2017, Feb. 13). "When Michael Crichton Reigned over Pop Culture, from ER to Jurassic Park." *Vanity Fair*. Retrieved from https://www.vanityfair.com/hollywood/2017/02/michael-crichton-reign-over-pop-culture-jurassic-park-westworld

discussions seem rather patriarchal in our more sensitive age today, but *State of Fear* seems to have special relevance for today. While this novel received almost universally negative reviews, its look at the politics of climate change resonates even today during election season in the United States.

BILL & TED'S EXCELLENT ADVENTURE (1989)

What can possibly go wrong when a couple of high school kids get their hands on a time machine? The adorable slackers Bill (Alex Winter) and Ted (Keanu Reeves) travel back through the ages for a school project, creating chaos in the twentieth century. While the movie has spawned two sequels and several spinoffs with the original characters, here we're going to focus on just the first one, the one that people tend to remember best.

The Science of Time Travel

Ted and Bill are the high school kids that teachers hated, but who probably grew up to have the most interesting lives, to be honest. But when we see them, they're a couple of people who spend far more time working for their band (the Wyld Stallyns) than studying. What's worse is Ted's dad is the police chief of the real-life community San Dimas, California. As the deadline for a major history report approaches, Ted's dad threatens Ted with military school if the teenager gets a failing grade.

As it turns out, history wants these kids to succeed. It's because in the year 2688 (seven-hundred years from the present), humanity has transformed into a peace-loving species due to the very music of the Wyld Stallyns. The leaders send Rufus (George Carlin) back in time in a time machine that looks suspiciously like a phone booth. (Note that unlike the TARDIS police box, this booth does not look bigger on the inside than the outside.) After some confusion about whether this time machine is real or not, Bill and Ted hop in the phone booth and steal Napoleon Bonaparte from the eighteenth century before deciding to visit other historical ages and grab their most famous characters.

How does this time machine actually work? Unfortunately for scientists, but very fortunately for audiences, this film is a comedy. This means that detailed scientific explanations aren't of much help, unless you're trying to build up to a punchline. So we never do hear about the mechanics of Ted and Bill's phone booth, although we do get a few clues from their time-traveling situation. Every time the teenagers (often accompanied by historical figures) jump into the booth, an antenna on the top directs their direction through time and space. (As you find out in the film, that antenna is a crucial element for how time travel works—but we'll let you learn the consequences, instead of spoiling them here.) The booth (from an outsider's perspective) appears to fall into the ground, and then it gets shot through a technicolor wormhole before emerging on the other side. This means the movie uses a familiar trope so that viewers aren't too distracted by the mechanics of time travel. Instead, people can focus on the comedy and the plot. And since we're already familiar with wormholes from *Doctor Who* (an obvious predecessor series to Bill & Ted), we can move right along to the consequences of their time travel.

Here's an interesting thing about the movie, it doesn't go into the usual tropes of the grandfather paradox or alternate futures or any of the other time-travel plots that we've already explored. In fact, this approach might be more scientifically rigorous, as some experts argue that once an event has "happened" in the timeline, it cannot be altered. This is best shown in the famous line that Rufus delivers, when giving a primer on how to time travel: "No matter what you do, no matter where you go, that clock, the clock in San Dimas, is always running." So the teenagers have a deadline to meet and know that their own "time" is not going to pause for them while they're checking out girls in the Middle Ages, for example. But the characters, clumsy as they are in history class, are actually creative problem-solvers and scientists. They work out by themselves how to (for example) get a key to unlock a door in one of the places they are visiting. How? By taking advantage of that causal loop that we've talked about before. Here it is in action:

Bill: "Can we get your dad's keys?"

Ted: "We could steal them, but he lost them two days ago."

Bill: "If only we could go back in time to when he had them and steal them then."

Ted: "Well . . . why can't we?"

Bill: "Because we don't got time!"

That's an interesting observation on Bill's part, by the way, "We don't got time!" They know that they are living out a timeline that is going to keep going forward, and they can't keep traveling forever while they have a history report that is due. But Ted jumps in with an easy solution:

Ted: "We could do it after the report!"

Bill: "Ted, good thinking dude! After the report, we'll time travel back to two days ago, steal your dad's keys, and leave them here!"

Just as Ted and Bill realize the solution, they then discover the magic of causal loops—Bill suggests they hide the keys behind a nearby sign, and then discovers the keys are *already there*. So clearly, their future selves did do the deed. Bill and Ted promise that they are going to do the same thing, but we don't see it on screen. So either they broke the causal loop continuum, or it's assumed that it happened and viewers don't need to witness it.

One last thing about causal loops and timelines, I was convinced the entire film that stealing somebody like Napoleon Bonaparte would irrevocably alter the timeline. If the famous French general had known about nuclear weapons, for example, how would that have changed his fighting style? But when he arrives in San Dimas, oddly he's not in a war-making mood, he hangs out at the local shopping mall and the swimming pool. Perhaps he sees the whole thing as a vacation. But the timeline of *Bill & Ted* carries on with no alterations, and no strange knowledge gained by any of the historical personas they gather. I guess time is indeed very hard to change.

The Science of Education

Originally I was going to write this section based on the "Dale's cone" model of education that claims that people remember 10 percent of what they read, 20 percent of what they hear, etc. But a little more digging reveals that the cone model is actually *not true*. While Edgar Dale did provide a cone of learning in 1946, it didn't include percentages and he also told readers that it was not a firmed-up model.[1] So it goes to show why you always need to check your facts. That being said, ever since discovering that, I've included more "experiential learning" in my lessons, meaning that I try to bind the theoretical knowledge I teach students with real-life experience. From what I can tell, they seem to enjoy it. My personal favorite was working with a class of aviation students and having them do reviews of a popular Canadian television show that featured aviators. I won't mention the name of that show here, as the students had almost universally negative reviews based on the maintenance shown in the show. But the point was, they were engaged.

What does this have to do with Bill & Ted? The students are told to write a history report, but not only do they need to write this report, they need to present it on stage. Now, I remember being a student and this kind of group work drove me crazy for two reasons: (a) I've always enjoyed writing more than performing and (b) I had to depend on a team to do the assignment, which was difficult for me as I enjoy having things under my sole control. But as a teacher, I've come to realize that I was the oddball. I love writing and I love reading, but many students—especially students brought up with YouTube videos and Snapchat and other technology not available to

1 Thalheimer, Will. (2017). "People remember 10%, 20% ... Oh Really?" Work-Learning Research. Retrieved from https://www.worklearning.com/2006/05/01/people_remember/

me in say, the 1990s—prefer using other ways to get their information. I can't promise that the students necessarily remember more from researching things on YouTube and doing quick in-seat presentations with the class, but (as much as possible as they can from assigned work) everybody does seem to get involved and engaged.

Lessons learned from this experience include always checking your sources, but also, making sure that when you are trying to teach somebody something that you *know your audience.* You are going to present information differently to a class of eight-year-olds than to a group of adult executives, right? Think of that whenever you build a presentation for work, whenever you write a memo for your boss, or whenever you have to do a history assignment. Create your creation with a clear end-goal in mind, picturing who will be consuming that information, and why. Nobody needs a retention chart to tell you that if your audience is interested in the information, they are more likely to remember it. Perhaps the best way is by writing, or perhaps it's by doing a skit. Or maybe you pick a few methods all at once. Whatever it is, make sure to think of others' needs before your own, and you'll do just fine.

The Legacy of *Bill & Ted's Excellent Adventure*

The legacy of *Bill & Ted's Excellent Adventure* lives on in an Internet meme—even though the Internet was in a primitive state back then compared to today, and none of us could imagine the powerful personal computers we hold in our pockets and on our wristbands. Specifically, *Wired* points out, an Internet meme called "Conspiracy Keanu" is based on an image of him playing Ted. By 2014, the magazine continued, Bill and Ted had inspired video games, comics, cereal, and a musical.[2] In fact, costar Alex

---

2 Watercutter, Angela. (2014, Feb. 7). Bill & Ted At 25: Dude, Bet You Didn't Know These 7 Gnarly Facts. *Wired*. Retrieved from https://www.wired.com/2014/02/bill-ted -25th-anniversary-trivia/

Winter told *Wired* that he feels the story anticipated the arrival of the Internet. "In the same way that *Doctor Who* has had such a massive resurgence because of Internet culture, I think that *Bill & Ted* absolutely has a connectivity with Internet culture," Winter says. "Because it's all about these guys who can go wherever they want really fast, with ease, which is sort of like what the Internet is." (And also, although Winter didn't say this directly, Bill and Ted rely on entering numbers into a phone booth and watching a result spit out—which is something similar to how many students research their essays today.)

There's perhaps no better symbol than Bill & Ted's enduring legacy, though, than the fact that the series will have yet another sequel in 2020. Called *Bill & Ted 3: Face the Music*, it will be the first movie of the franchise since 1991. That's an incredible span of time and speaks to a lot of confidence, on the part of creators, that Bill and Ted still have relevance among an eighteen- to thirty-five-year-old audience who were either small children or not born when the first two movies came out. That said, the franchise appears to be banking on nostalgia from people who are a little older, since Bill and Ted are portrayed as bumbling dads who are still waiting for that big hit.

Bill and Ted have us all questioning why it's important to study history. The thing that most teachers throw around is that it helps us avoid making the mistakes of the past once again, but there is a more compelling argument than that. History will help us make multidisciplinary decisions in fields that demand them. The book *The Mosaic Principle* (Nick Lovegrove) argues that the most successful people develop a deep understanding of a topic or two and then apply them across a broad set of spheres—public, private, educational, and so on. History helps us make the connections and bring broad groups of people together to achieve shared goals.

So while on the surface *Bill & Ted's Excellent Adventure* looks like another silly time-travel movie, by movie's end the teenagers realize that the best projects do come through a group of interested people working to a goal that will benefit greater society. In their case, it's creating the most memorable history project ever. But in today's interconnected world, this could mean creating better resources to preserve our water, our land, and our environment in general from external threats. Or it could mean building a society with more provision for diversity than past ones. That world will be one that you can help shape, I hope, so think about what you'd like and then build your career to try to attack the problem.

OUTLANDER (1991–PRESENT)

Outlander is one of those franchises that is so far-reaching that it's hard to capture in a short text. The original novels and works by Diana Gabaldon have sprawled into a larger universe, including a musical, a graphic novel, and (perhaps most famously) a drama series. While our examination of time travel will focus on the world of the novels, other parts of this analysis will discuss the wider universe.

The Science of Time Travel

How does time travel work in *Outlander*? Let's consult one of the fan wikis to know more, since these people live and breathe the universe of *Outlander*. The series follows the adventures of Claire Randall, a British Army nurse of the year 1946. She and her husband, Frank, have been separated due to the previous war years and they decide to go on a second honeymoon to Inverness, Scotland. Claire goes alone to some standing stones (a sort of upright stone common in older cultures) when she hears a buzzing noise, but upon touching one she faints and wakes up in the year 1743. Claire is called an "outlander" by the locals because she clearly is not from around there, and she struggles to fit into a culture that is some two hundred years before her own.

In the series, according to the fan wiki,[1] it appears that time travel (as far as people can tell) is related to genetics. For example, both Claire and her daughter, Brianna, have the ability to travel through time. One traveler, the Comte St. Germain, specifically looks for

---

1   Outlander Wiki. (2019). "Time travel." Retrieved from https://outlander.fandom.com /wiki/Time_Travel

people who have a "blue aura" about them. It's said to be a trait of time travelers, although little is known about what that means.

Time travel also seems to be closely tied to particular locations in the series, such as Craigh na Dun, a stone circle near Hadrian's Wall and a tunnel under Loch Errochty. Strangely, there also seem to be only particular times that the portals open, such as around the summer and winter solstice. (Claire's particular time-travel day is May 1, also known as "May Day" or "Beltane," a major Gaelic festival. And yes, other time travelers took their journeys on that date.)

While the novel and the show seem to shy away from the mechanics of time travel (which makes scientific analysis for us really difficult), at the least they try to establish clear rules, according to IGN.[2] Time passes at the same rate regardless of location, gemstones are required to move through time, time travel is most likely to occur at specific locations and dates (which we've already discussed here), that it's genetic, and that—most tragically—there are only so many opportunities to go through time. IGN explains on this last point:

> Time travel on *Outlander* isn't as simple as stepping through a portal; instead, it's an exhaustive experience that wears on the mind and body. As such, people can't sustain multiple trips back and forth, which becomes important as Claire weighs the decision between whether she should return back in time to reunite with the love of her life, or stay with her daughter in the 60s and live out their lives together.

An interesting thing to ponder is how Claire's presence in the past will affect the future. As stated earlier, she is a nurse. And because she is a nurse, she feels duty-bound to help people. By

---

2   Schwartz, Terri. (2017, Sept. 7). "The Rules of Time Travel According to Outlander." IGN. Retrieved from https://ca.ign.com/articles/2017/09/07/the -rules-of-time-travel-according-to-outlander

helping people, points out CinemaBlend, she could be very much altering the future.[3]

> Then, there's the fact that Claire was saving lives all over the place once she ended up in the past. In fact, if she hadn't turned up when she did, Jamie himself would have died of his wounds. If the lives she saved were those of people meant to die in the eighteenth century, she was altering the future in some huge ways. Some people even died because of her, and perhaps they were people who were meant to live long lives that contributed to history in the original timeline. We just didn't know, and it was possible *Outlander* was setting us up for a rude awakening at some point in the future.

So do Claire's actions in the past have an effect upon the future that we should worry about, if we were living in the same timelines? CinemaBlend denies that this is creating paradoxes, arguing that Claire's contributions could not have measurably altered the future since (based on spoilers I won't reveal here) they could not trace any measurable impact. "For those of us who are sticklers for rules when it comes to time travel," CinemaBlend said, "it should be a relief to stop worrying that Claire's actions will result in something horrible happening in the present. Her meddling in time already happened and resulted in the timeline she knew before she came to the past, so there's nothing she can do to ruin the future."

The Science of Feminism
Outlander's (television) depiction of feminism has come across many points of view about its success. An *Atlantic* article from 2014

---

3   Hurley, Laura. (2017). "Outlander May Have Just Settled A Huge Time Travel Question." CinemaBlend. Retrieved from https://www.cinemablend.com/television/1701399 /outlander-may-have-just-settled-a-huge-time-travel-question

captures the praise and criticism of the time, although to be sure the series may have done better in later seasons. One writer, Julie Beck, particularly praises the character of Claire, saying she is a stronger female character than many other fantasy people of that era, including those of Game of Thrones.[4] "She's strong, but not in the lazy way 'strong female characters' are often written, when a writer takes a typical macho action hero, and gives the character another X chromosome,'" Beck said. "She owns her sexuality, but isn't sexualized. She's smart, caring, and has some creative swears. ('Jesus H. Roosevelt Christ!')"

But another *Atlantic* writer taking part in the same conversation, Olga Khazan, criticized the series for going too far down the "Fifty Shades of Plaid" route—that is, showing gratuitous sex that is not advancing the plot very much. Claire also faces negative experiences, such as a near-rape by her husband's ancestor. "Claire does seem extremely progressive, and that's actually pretty historically accurate. Right after World War II, women had just taken on all of these traditionally male roles, and so you had this slight feminist surge for a few years," Khazan said. She continues:

> However, I do worry that she's maybe a little too—dare I say—pushy. Like, common sense might dictate that if you find yourself transported through time to what appears to be an eighteenth-century Scottish castle, you don't talk up your own mysterious healing powers, take an acute interest in the shoulder health of the low-ranking stable boy, or debate rape culture with the Scottish Laird who has the power to take you prisoner.

4   Beck, Julie, Green, Emma and Khazan, Olga. "Outlander: False Feminism?" *The Atlantic.* Retrieved from https://www.theatlantic.com/entertainment/archive/2014/08/outlander-reviewed-is-starzs-new-show-actually-feminist-or-pandering-to-feminists/378637/

So, how to reconcile these points of view? Far be it from me to tell the TV writers how to write their script. Instead, I would point to people who promote feminism and who are experts in doing so. There are many people that you can turn to (Gloria Steinem, Dorothy Pitman Hughes, Betty Friedan, and the like). Here, let's briefly quote an organization that does its best to promote multicultural feminism: the Association for Women's Rights in Development. Here, in their words, are how they try to promote their perspective all over the world:[5]

- Fostering community and sharing information through the Young Feminist Wire (a publication of theirs that is supposed to "share information" and "build capacity," which includes "community building."
- Researching and building knowledge on young feminist activism.
- Promoting more effective multigenerational organizing.
- Supporting young feminists to engage in global development processes.
- Collaboration . . . to ensure young feminists' key contributions, perspectives, needs, and activism are reflected in debates, policies, and programs affecting them.

This is a theme I'll touch on again and again in this book, but the best way to portray a community accurately (in fiction or in discussions about it in real life) is to directly involve it. This includes all ages and genders and minorities in a community. That way, the perspective received is as diverse as possible. This policy of inclusion works well in all fields of life, even in matters such as science. Too much of history has been done by a single type of

5   Association for Women's Rights in Development. (2019). "Young Feminist Activism." Retrieved from https://www.awid.org/special-focus-sections/young-feminist-activism

person. Imagine the insights that we will receive if we bring in more people, from all walks of life, in solving our biggest societal problems.

The Legacy of *Outlander*

Perhaps the best thing about *Outlander* is how much of a benefit, in some ways, that the series has been to Scotland. While there is a caveat with putting increased attention on one part of the world (in terms of the effects that thousands of new tourists suddenly arriving can have, such as homogenizing the culture or overwhelming the infrastructure of destinations abroad), there is certainly an interest that is encouraging people to learn more about Scotland's history and peoples.

One interesting offshoot of *Outlander* is a resurgence of interest in the dialects spoken in Scotland, although one writer in The Conversation warns that what you see on TV is not true "Scots." Rather, it's "a stage-Scots—essentially English dressed in tartan and cockade—yet it is still to be cheered." According to The Conversation,[6] Scots is (still!) spoken across the country, most especially in rural areas. It actually has about 1.6 million speakers in Scotland, using "loan words" from Germanic dialects, Gaelic, Dutch, French, and Norwegian. Its presence in *Outlander* is something to be applauded, the piece continues, because "this language has been repressed—forced out of education, media, and business life." The history behind this repression cannot be captured in a short summary, but essentially it arises from events after the Act of Union (which brought together the governments of England and Scotland), which was signed in 1707. But the encouraging fact is this language continues to be spoken—and is now heavily

---

6   Watt, Irene. (n.d.) "Outlander is boosting a renaissance of the Scots language—here's how." The Conversation. Retrieved from https://theconversation.com/outlander -is-boosting-a-renaissance-of-the-scots-language-heres-how-101643

promoted—and that there's a resurgence of interest. It would be nice if people were as enthusiastic learning it as fictional languages such as Dothraki (*Game of Thrones*), Klingon (*Star Trek*) and the Elvish dialects from *The Lord of the Rings*. "The past for Scots may have been bleak, but the present is improving and the future is starting to look optimistic. *Outlander* is part of the rehabilitation, putting Scots back on the lips of our ancestors where it belongs, in its central place in Scottish identity and heritage," the piece concluded.

And the language is not the only thing of Scotland that *Outlander* sparked an interest in. According to a 2016 article in the *Globe and Mail*, the series "has sparked a craze for all things Scottish on the catwalk and beyond" to an extent that no other series of novels has done since Sir Walter Scott's *Waverley* series in the nineteenth century. The newspaper observed tartans on fashion runways in the big cities (Paris, London, New York), and in-home décor (Ralph Lauren, Isaac Mizrahi). Obviously, these trends are aimed at more high-income people, but a wider range of participants can fashion their own: "Internet sites such as Etsy and Pinterest are bulging at the seams with DIY knitting projects created by fans of *Outlander* wanting to imitate Claire's romantically rustic look," the newspaper adds.[7]

And there's also the tourism effect on Scotland, with more people coming to the Highlands since 2014, the *Globe* added. "The biggest surge is at Doune castle near Stirling, where much of season one was filmed. Visitors to the site—fictional Castle Leoch—increased 44 percent to 32,540 people during three record breaking months last summer." Again, we must be careful about getting too excited—the success of these initiatives depends on the

---

7   Kelly, Dierdre. (2016, March 17). How Outlander has made tartan the new black. *The Globe and Mail*. Retrieved from https://www.theglobeandmail.com/life/how-outlander-has-made-tartan-the-new-black/article29272276/

enthusiasm of the peoples that it depicts—but in an overall and vague sense, any series that encourages people to learn more about history or other cultures (even the fictional ones) is a good thing. It is through connecting with others that we can make a difference in our world, even in small ways.

GROUNDHOG DAY (1993)

This classic Bill Murray comedy, on the surface, looks like a simple story: What happens when you relive the same day over and over again? And how do you get out of what appears to be a terrible, never-ending time loop? But underneath, there is a more compelling story about character growth and about mastery, which this movie explores well in its "lessons learned" for viewers watching the show.

A Special Note

Connors goes through a very dark period in the film and falls into depression, including suicide attempts. While the movie treats depression as a joking matter, today we know that depression is a serious, yet treatable condition. There are many free services available if you are depressed, if you are suicidal, or even if you're not sure. If you are in crisis and in the United States, please call 9-1-1, free National Suicide Prevention Lifeline at 1-800-273-TALK, or text the Crisis Text Line (text HELLO to 741741). There are similar services in just about every country imaginable, so please seek them out. Remember: you are worth it, and we believe in you.

The Science of Time Travel

The premise of *Groundhog Day* is ridiculously simple, until you start to think about how he time travels. So, our friend Phil Connors is a TV weather forecaster and he has been sent to a small town in Pennsylvania for "Groundhog Day," an early February event where a groundhog, according to mythological lore, looks for their shadow. Based on what they see, we're either going to have a short winter or a long one. Connors is not thrilled to be in a small town and is especially not thrilled about having to cover an event involving a groundhog. He's rude with his coworkers and repeatedly says things that show he thinks highly of himself and very low of others.

The next morning, he wakes up in his hotel and gradually realizes it's the same day. The radio plays the same music, the people he meets in the hotel and on the street say the same things, his shoes get ruined in the same puddle he stepped in yesterday, and so on. Connors gets through this strange repeat and makes it to bed once more, only to find himself stuck in a time loop. He gets arrested? Ends up in the same hotel bed when the clock resets. He dies? Ends up in the same hotel bed in the morning (we hesitate to say "the next" morning, since it's the same time loop over and over again).

So, what exactly is a time loop? One of the first people to think about this problem was Kurt Gödel, a literal contemporary of Einstein (in the sense that the two geniuses worked together at Princeton's Institute for Advanced Study in the 1940s). While a time loop, at first, doesn't sound like an intuitive thing, Gödel found an explanation. "If the universe were rotating, it would then be possible for an object to travel in a certain closed loop in space and come back to its starting point before it left! In other words, a person could travel around a loop in space—and discover that it is also a loop in time," said theoretical physicist Paul Davies in a conversation with the online publication *Edge*.[1] A related concept to a time loop is something called "The Big Crunch." Here's how it works. As far as we understand it, the universe began 13.8 billion years ago in an event known as the Big Bang. Space and time (as we know it, at least) simultaneously came into being. Space began expanding, and as time went on it actually began accelerating that expansion for reasons we are still trying to figure out. And time marched on,

---

1   Edge. (n.d.) Time Loops: A Talk With Paul Davies. Retrieved from https://www.edge
    .org/conversation/paul_davies-time-loops

eon after eon. If the universe began, most likely there will be an end to it. The Big Crunch suggests that if there is enough mass in the universe, at some point we will reach an end point in the expansion, just like stretching an elastic band. (The metaphor isn't perfect, of course, as one can break an elastic band—but you get the idea, I hope.) Anyway, once you reach the end of the expansion, there will be a rebound where the universe collapses in on itself. This could, some physicists suggest, make time run backward. So time itself may run in one big loop that has forward motion for a few billion years, and then backward motion for the last few billion years.

Understand, though, that The Big Crunch is unlikely to come to be. Scientists have estimated the amount of mass in the universe and they think, even including the dark matter and dark energy we cannot directly see or account for (at least yet), there isn't enough mass available to make the universe rebound. Instead, we face a long, cold, dark end. Stars will grow farther and farther apart, dispersing galaxies will run out of the gas and dust they need to fuel new star growth, and gradually everything will just run out of fuel and dissolve. Luckily, such a fate is many billions of years beyond our lifespan today, so it's not an imminent problem. But certainly something we may want to see if we can fix, if we're planning on staying alive as a species indefinitely.

These are big thoughts. And that's the thing about science— think about a problem long enough, and you literally find yourself pondering the fate of the universe. While every scientist has a very small problem that they are an expert in, working together with other disciplines they do solve very big problems and come up with very interesting discoveries. Just in the past few months before I wrote this manuscript, an international team with many telescopes imaged a black hole for the very first time. Weeks later, I attended an elementary school class and the students demanded to see pictures of it; to them, it was normal and taken for granted. But for science journalists and scientists? It still blows our minds.

The Science of Mastery

Connors eventually settles into his small-town life and begins to use his extra time to learn new things. For example, we see him sign up for piano lessons and begin playing his first few tentative notes with a teacher. Fast-forward somewhat, and Connors suddenly has rocket-finger hands flying across the keys. We know that there is a long period of practice and study and frustration in between. So how do you get from a beginner state to an advanced one? The answer, as best as I can tell from the literature, is a concept called "deliberate practice." One of the modern pioneers of this idea is K. Anders Ericsson, a psychology professor at Florida State University. You may not have heard his name, but you probably have heard of the "10,000-hour rule." That's the rule, popularized in books by Malcolm Gladwell and countless other people, that all it takes to be good at something is to spend 10,000 hours doing it. That's a simplification, though. As multiple interviews with Ericsson point out, what you need to do is 10,000 hours of *deliberate practice* to become a master. This doesn't mean just playing through a piano piece, but doing stuff like picking through scales, figuring out why your fingers keep slipping through a fast passage, and constantly reworking your technique to make the music flow to listeners' ears. Every sport or instrument or discipline has its own metrics for mastery, and you need to focus on the hard parts to reach your 10,000 hours. That's why professional teams do drills. That's why writers sign up for boot camps. That's why athletes have fitness trainers, to point out their weaknesses and fix them.

I encountered this situation myself when learning Russian for a trip I wanted to take. Russian, if you're an English speaker and not familiar with it, has a completely different language structure. There are six different cases of words that are used, and you need to be very aware of those cases because the endings of words change depending on what case you are using. And believe me, those cases are not intuitive. I spent years trying to remember the different cases in dative and genitive and so on. After more than four long years of study, I began to feel a difference—I was carrying on conversations. Then I found myself in Russia and Kazakhstan after about five years later. I certainly was nowhere near fluent, but I could walk into a store and find groceries. I could do simple conversations with passersby ("Where is . . .," "Excuse me," and so on) and understand the responses just fine. And by the way, in case you've heard the myth that only children can learn languages, I started this process in my early thirties and made significant progress. I didn't learn as fast as a kid or teenager, but I made it far.

Here's another thing to know about mastery. Just because you know how to do a thing well, doesn't mean you need to continue doing it. After careful consideration, I dropped my Russian lessons after my second and last trip to the region. I had accomplished my goal (watching a space launch in the region) and had new goals in front of me, including finishing a PhD and writing this book. Being a master means knowing how to deploy your limited resources and time to accomplish a goal. Knowing that I needed to earn a living, I stopped Russian and began working on new projects. But who knows? Maybe in some distant year or decade, I'll have the willingness to restart again.

The Legacy of *Groundhog Day*

One modern reviewer brilliantly calls *Groundhog Day* "a slow-burn existential nightmare" on top of being a slow-burn comedy. "The movie's genius lies in blending the two seamlessly without making much of a fuss about either one," Josh Larsen wrote on his website.[2] He smartly points out that at first, the movie is a satire of living in a small town. People come out and do the same thing, year after year; some people love that structure, while some people

2 Larsen, Josh. (2017). *Groundhog Day*. Larsen on Film. Retrieved from http://www .larsenonfilm.com/groundhog-day

really, really need to escape it. Connors starts out as the second type, using Wes Anderson–like antics to try and break out of the small town, Larsen says. But as the movie moves on, we begin to see greater themes. For Larsen, it's moving away from selfishness: "This shift from selfishness to selflessness lies at the heart of many world religions, from the Zen Buddhism Ramis had embraced to orthodox Christianity."

But there is also the theme, literally speaking, of resurrection and renewal. Any student of mythology has likely heard of The Hero's Journey, pioneered by Joseph Campbell and used lovingly by so many people (including George Lucas, who made a little franchise called *Star Wars*). Without going too deep into the journey, understand that many of these heroes do have to go into the underworld, to confront death, sometimes literally the personage of death, and then come out again, equipped with new knowledge. In *Star Wars*, Luke Skywalker retreats into a forest with Yoda and emerges with most of the marks of a Jedi warrior. In *Game of Thrones*, Daenerys Targaryen trusts her husband's and unborn son's health to a seer, and emerges from a horrific experience thinking that she can only rely on herself. In ancient mythology, Odysseus visits the underworld on his decade-long journey home, to chat with friends and elders long dead and to learn how to get ready for his return to family. And yes, we see the same thing in *Groundhog Day*, except exaggerated to comedic effect. When Connors literally sees the same day over and over again, he moves through a depressive period to one of great productivity, when he's able to accept what is going on around him and to use his circumstances to the greater good. This isn't, by the way, a statement on treatment for mental health, which is much more serious than how a comedy portrays it. But it is a reflection that by going through difficult experiences, in some instances we grow. We become stronger. And then we can turn that around and help others with the knowledge we acquired.

To be sure, *Groundhog Day* is locked in many senses to the 1990s. I mean, who watches the weather forecast anymore? Most of us do a quick check on our smartphones before moving on, although during natural disasters we are more likely to seek a TV broadcast (albeit, online). The people in the small town are isolated, in a sense that is hardly true anymore since the Internet literally brings the world to our doorstep—except in the most rural areas. For these reasons, it may be hard to have *Groundhog Day* connect to modern audiences. Yet it is a facet of modern culture. We still hear people joking that today "felt like Groundhog Day," in that like Sisyphus, we were rolling a stone (or work, or something else) up a hill only to watch it roll all the way back down again. We've all had those days. And for the most part, Connors's experience still holds up to us—even if our weather forecasts come through phone apps rather than through live events.

PART IV
INTO THE
MODERN ERA

AUSTIN POWERS: THE SPY WHO SHAGGED ME (1999)

Raunchy, irreverent, and making fun of the best parts of James Bond movies, the *Austin Powers* franchise defined much of spy spoofing in the early 2000s. With both good (Austin Powers) and evil (Dr. Evil) being played by Mike Myers, this literally showed a two-sided personality approach to a common problem in science fiction: how to deal with time-travel paradoxes. So let's dig in to this second of three films about Austin Powers (and the only one to focus on time travel).

The Science of Time Travel

Austin (Mike Myers) is on honeymoon with his wife Vanessa (Elizabeth Hurley) during the present day, 1999. But their wedded bliss is cut short when Austin discovers that his newlywed partner is actually a "fembot" with ulterior motives. It's just the first of several pranks from Dr. Evil (also played by Mike Myers), who isn't a big fan of the spy. While Austin is distracted by Dr. Evil's first prank, the villain is working on a more elaborate plan—to create a time machine, go back to 1969, and make it more difficult for Austin to have sex. Once Austin finds this out, the Ministry of Defence sends him back in time in, of all things, a Volkswagen New Beetle—a popular car of the 1990s that was supposed to evoke nostalgia for the original Volkswagen Beetle of the 1960s. There, Austin must search for the stuff that Dr. Evil stole—although, as it turns out, there may be an alternative way to get what he really wants. We'll let you watch the film to learn more (with the requisite parental caution advised for mature themes and all that).

Being a comedy and not a science movie, we don't get to see much of how time travel works in the Volkswagen. It appears to borrow a bit from *Back to the Future*, where you hit the date you want and then drive super-fast. At some point within the design limit of the Volkswagen's speed, physics takes over and you find yourself at the time that you want. But what's really interesting is that once a character arrives in the past, if he or she was born then, they revert (in a sense) to their younger selves. For example, any teeth fillings or alterations to your body are removed. We also learn that the Volkswagen appears to be a little more flexible in time travel than Dr. Evil's machine. Evil can only go back in time in his own lair, while the Volkswagen can cruise across both time *and* space. But unlike *Back to the Future*, the Volkswagen cannot go to a specific time of day, which would have been a terrible thing if they had to arrive in time for something like a lightning storm. Rather, you need to be flexible with your timing.

From here, any science that we could talk about it is fairly theoretical. Since the comedy doesn't go into great detail about how they work, and since this isn't meant to be a "hard" time-travel story, you need to sit back and enjoy the plot rather than try to pick apart all the problems this story creates. "All of them run into serious paradoxical and logic issues and if you come across [a time machine], it is best to not think about it too much and just enjoy yourself," the blog *Robot Plunger* suggests.[1] One of the most difficult parts to accept is what happens when Austin runs into a younger version of himself, for example. Without going into too much detail, the movie avoids the usual paradox of characters possibly accidentally annihilating each other upon sight or upon touch. Rather, the Austins are able to actually team up with each other to solve problems or just to have some fun together. Even weirder, the

---

1   *Robot Plunger*. (2011, Aug. 24). "The Best Ways To Travel Through Time." Retrieved from http://www.robotplunger.com/2011/08/the-best-ways-to-travel-through-time/

two Austins seem to get along just fine in the same timeline—and other characters accept their presence. Another confusing thing is when Austin's ability to have sex disappears in 1969, you would think that it would affect his abilities all the way along the time. But the movie only shows Austin having difficulties in the present day (1999) and makes no reference to anything having changed in the past. Both due to irrevocable paradox issues and due to decency issues, we'll not dwell too long on how to solve this one.

Some fans have also pointed out that at least one of the characters (Dr. Evil's son Scott, played by Seth Green) appeared too young in 1999 to have been conceived in 1969. While Green was technically born in 1974, some online forums complain that he was acting more like a person in his twenties than somebody who was thirty. We suppose it depends on the thirty-year-old, but some of the conspiracies out there suggest that perhaps Scott's growth was somehow decelerated as an off-screen joke. (And to be fair, actors are often asked to play people older or younger than themselves—particularly in that twenty to thirty sphere, since studios prefer working with that age group rather than teenagers due to workplace regulations about children and hours worked in a day.)

The Science of Volkswagen Beetles

Here's a fun paradox to consider: Could the Volkswagen New Beetle (a car from the 1990s) really fool people in the 1960s? Perhaps the screenwriters were taking inspiration from *Doctor Who*'s Time And Relative Dimension In Space (TARDIS) machine, which appears like a police phone-call box on the outside and on the inside is very clearly a time machine. But here, we have to make a couple of assumptions about people.

Before we get into the design, understand that the VW Beetle has a somewhat dark history. While designed by German designer Ferdinand Porsche, it came as a result of Nazi party leader Adolf Hitler's dream to create an affordable car for German families.

Hitler himself stole some of the ideas from inventor Josef Ganz, who put a prototype car in a show in 1933. Ganz was Jewish and had to close his business and flee to Switzerland as anti-Jewish activity ramped up in Germany. (To be clear, the Nazis never released this car on the mass market, so there is a clear distinction between what Hitler wanted and what people were actually driving in the decades since.)[2] Beetle evolved and changed over the decades, and we see this between the car of the 1960s and the car of the 1990s. Even when the New Beetle came into mass production in 1999, it was easy to spot the difference based on the caliber of the car alone—specifically, they were in much better shape than most vehicles from the 1960s and 1970s. This was clear especially in my city, Ottawa, whose harsh winters can easily destroy vehicles through rust and general wear and tear unless you have a dry place to store your car and you're prepared to clean it regularly.

Some car fans will notice weird things about the New Beetle, whose engine was in the front (not the back) and which had front-wheel drive. This means that the car will drive slightly differently and sound different. But all of this discussion about driving styles and engine locations forgets to mention the very obvious—the two cars look very different even to a casual observer. The newer Beetle has smoother lines on the front and the older Beetle has defined wheel-wells. The trunk shape is more pronounced on the old, and the headlights have different shapes as well. And we're not even mentioning the material problem—the fact that the 1999 version would have used improvements in plastics and metals and other things that have represented the science of progress over the decades.

So it's clear that the *Austin Powers* franchise is having a bit of a joke with us, that it's very hard to make a car from 1999 pass for

---

2   Threewitt, Cherise. (2019). "Did the Nazis invent the Volkswagen Beetle?" How Stuff Works. Retrieved from https://auto.howstuffworks.com/under-the-hood/auto-manufacturing/did-the-nazis-invent-the-volkswagen-beetle.htm

a car from 1969. Perhaps the larger lesson of the joke is that what is old is often new again. These days we're seeing a resurgence in movies and TV shows celebrating the 1980s: *Stranger Things*, *Ready Player One*, the new *Top Gun* and the like. The Beetle had appeal in 1969 and had appeal again in 1999, although these days it's (at least temporarily) at the end of its nostalgia run. Pending a possible new electric car vehicle, the third-generation Beetle stopped being manufactured in July 2019.[3]

The Legacy of *Austin Powers*

At the time of *Austin Powers*'s release (this was in fact the second of the three films), it seemed you couldn't go anywhere without hearing about the film. Everybody was going about in Halloween costumes and quoting lines, as *Collider* revealed, but in 2017 the ardor had cooled.[4] "Certain lines are half-heartedly quoted from time to time, but by and large it's not an evergreen IP. In 2017, cultural enthusiasm surrounding *Austin Powers* feels as dated as being nostalgic for Y2K or Ricky Martin," it wrote. *Collider* adds that the screenwriters played it smart with the *Austin Powers* franchise, going light on the pop-culture jokes and relying on timeless humor such as line-of-sight gags. And of course, Myers is an able comedian on his own and talent definitely spans the ages. Just ask any student of Charlie Chaplin.

It's unclear whether we'll see more action from Austin Powers. A fourth movie has been in rumors for a long time since the last film was released in 2002, but the death of supporting actor Verne Troyer (Mini-Me) could make another film difficult to achieve, director Jay Roach said—while clearly saying that the decision

---

3   Silvestro, Brian and Perkins, Chris. (2018, Sept. 13). Volkswagen Is Cancelling the Beetle in 2019. *Road and Track*. Retrieved from https://www.roadandtrack.com/new-cars /a19155284/volkswagen-beetle-discontinued/
4   Chitwood, Adam. (2017, May 2). 'Does 'Austin Powers' Hold Up 20 Years Later?' *Collider*. Retrieved from http://collider.com/austin-powers-does-it-hold-up/

would be ultimately up to Mike Myers.[5] Bond, meanwhile, is expected to have its twenty-fifth film in the franchise in 2020, so we'll see if it continues in a serious vein as Craig once again takes the acting helm.

It's possible that the legacy of *Austin Powers* may depend on its ability to deliver sequels. Right now, it seems to be more of a film that evokes nostalgia in people of a certain age, in their thirties or forties or older. To crack that lucrative eighteen to thirty-five group, *Austin Powers* will need to reinvent itself as a franchise that can bring in new elements of pop culture while still appealing to those who enjoyed the original trio. It's something that other films have certainly been able to do—look at *Star Wars* for example. The 1977–1983 smash hit trio is finally back in the minds of youngsters thanks to a sustained effort by Disney; as I write these words, anticipation is building on the Internet for *The Rise of Skywalker* thanks to the release of a new trailer. This trailer is all over the geeky magazines and fans are eagerly dissecting it moment by moment, while talking about what it could mean for the franchise as certain characters pass away and new ones take their place. Perhaps this points to a way that *Austin Powers* can continue, by handing off some of its key roles to new actors, in the same way that *Bond* and *Doctor Who* do. This keeps the series fresh and allows it to diversify its content in a way that would appeal to more audiences. Let's hope that this happens. Otherwise, this time-travel romp to 1969 will appear a dated homage to boomers that, literally, are today's kids' aging grandparents.

---

5   Sharf, Zack. (2019, July 30). "Austin Powers' Director Says Fourth Movie Would've Gone Deep on Verne Troyer's Mini-Me." IndieWire. Retrieved from https://www.indiewire .com/2019/07/austin-powers-4-jay-roach-verne-troyer-mini-me-1202161908/

FUTURAMA
(1999–2013)

Futurama, like so many of the great shows and movies we've portrayed in the book so far, begins with a time-travel situation. Philip J. Fry, who appears to be an unremarkable human being up to this point, is frozen for one thousand years and wakes up again in the thirty-first century. Because Fry wakes up in the future, he then finds a spot working at an interplanetary delivery company and meets a bunch of loveable society non-conformists that fans followed closely for dozens of episodes, including a cancellation and a revival. The series is a fun deep-dive into time travel in that several characters use it—but it can be confusing because there are so many different forms of time travel employed. So here's a quick overview of how it's done, with the understanding that to really appreciate all the time travel . . . it's best to do some weekend marathons to appreciate the series.

The Science of Time Travel

How do characters in *Futurama* travel through time? Creators Matt Groening and David Cohen did not intend to use the trope extensively at first, but over time (which happens with many concepts) time travel ended up being a commonplace event. But what makes this patchwork of ideas interesting is that there is no single way that people travel through time. So it's hard to do a scientific analysis of what is right and what is wrong, since there are so many different directions to tackle.

To borrow heavily from the fan site TheInfoSphere.org,[1] here is a list of at least some of the ways that time travel occurred in

---

1   TheInfoSphere.org. (2015, Aug. 25). Time travel. Retrieved from https://theinfosphere .org/Time_travel

Futurama, and how possible they may or may not be. Note that there are some time-travel types that are not very well described in the *Futurama* universe, including singularities, unseen time machines, and time warps, so we've only covered a selection of the time travel situations in *Futurama* below.

Suspended animation: In the *Futurama* universe, Cryo-Tubes let Fry go to the thirtieth century, among other examples, through suspended animation. Wait, you say, that doesn't sound like time travel at all. You're put in a cryogenic freeze and wait out your time to awaken, like Rip Van Winkle. Actually, though, you can argue that it is a form of chronological time travel, the fan site argues. "Chronological" is how we all travel through time—second by second, always forward. What suspended animation does is it freezes your perception of time, has you lie dormant, and then wakes you up again when some time has passed—making this a type of time travel. (It's an idea that fans of *Idiocracy* are familiar with, for example.) Is this possible? Well, animals do it all the time when they hibernate. While there are efforts to break down the hibernation process for humans, so far we haven't been successful. But that's not to say we could never do it.

Time skips and forward traveling: These are both variants of moving chronologically through time, although *Futurama* employs devices to make their different applications really obvious. Chronitons (purple particles that are unstable in the field of time) create time skips in the *Futurama* universe, making time move forward in unpredictable bursts. As for forward travel, Farnsworth does have a machine to allow that to happen—although, unfortunately, we never learn exactly how it works. So focusing on the time skips, the question is whether we can slow down or speed up time according to our needs. Time depends in part on our perception of how fast or slow things are going—just remember that time you're sitting in a boring class in school (in which time takes forever) or you're engaged in yet another round of Zelda at your computer (where time passes quickly). If you want to feel like you're living longer, some bloggers suggest that injecting novelty into your day[2] will make time pass slower. This might be why vacations feel so much longer sometimes—it's because every day is a new experience. A related concept to Chronitrons are time vortexes, which use Chronitrons to pop people into random places. We've already seen this discussion in time machines that can travel people accurately through time, but not in space.

Wormholes: We've covered this concept over and over again in this book, so no need to revisit old ground. Just know that it happens in the *Futurama* universe using the Time Tunneller and a couple of other devices. A related (but poorly explained) point to wormholes in *Futurama* is the Nexus Point, which is a space-time location only accessible to a particular person coming from another space-time location. I doubt the laws of physics can be bent as far as the Nexus Point, although it does make for good sci-fi fodder.

The Time Rifle: This creates so many paradoxes that I suspect most scientists would just throw up their hands. In the words of TheInfoSphere.org: "A time rifle somehow causes future and past incarnations of a person or object to be swapped. A broken rifle is able to bring multiple versions of a person into the time period in which the gun was fired. Anything that happens to the younger version will also have happened to the older versions."[3]

The Farnsworth Effect: This is another one that gets props for sheer creativity, but appears difficult to prove in physics. Specifically, this is how TheInfoSphere.org describes it: "Farnsworth has discovered that the fluid keeping heads alive in jars is composed of crystalline opal that traps that head in a temporal bubble, and if it is ingested, the person will travel back to the head's era for

---

2   McKay, Brett and McKay, Kate. (2014, Aug. 11). Be a Time Wizard: How to Slow Down and Speed Up Time. The Art of Manliness. Retrieved from https://www.artofmanliness.com/articles/be-a-time-wizard-how-to-slow-down-and-speed-up-time/

3   TheInfoSphere.org, *ibid.*

a limited time."[4] I mean, A+ for effort, but this use of temporal bubbles is something difficult to explain except by just accepting that it happened and going with it.

The Science of "Fitting In"

One of my favorite things about *Futurama* is the folks working at the interplanetary delivery company, who are, to say the least, not really fitting into the world of the future. Fry is perhaps a more understandable case because he came from the past, but what about this group of people born into this society that are struggling to get by? Why I like this is because it makes viewers think through matters of privilege and how they affect us all.

Society is no more than a collection of people, and these people can be more or less tolerant to those who are not considered "the norm," so many groups have had to fight for rights over the years to be accepted—women, people of different genders, people of color, immigrants, etc. This fight is still going on today, as we see with discussions about the US-Mexico border, a woman's right to choose, or same-sex marriage.

If you don't "fit in," there probably is a greater reason beyond yourself that you're not "fitting in." Maybe it's a high school clique centered around sports when really, what you prefer is watching *Futurama* or other science-fiction series. Maybe it's because you identify with a gender that is not fully protected in society. All of which to say, if you don't "fit in," there's nothing wrong with you—it's more that there's something wrong with the people around you. Fry had to find his place in a whole other century to figure things out. Your people are out there, you just may have to do a bit of soul-searching to track them down. But don't give up. Keep searching. You will be accepted as soon as you find your group.

---

4   TheInfoSphere.org, *ibid.*

The Legacy of *Futurama*

Webcomic writer Gabriele Cheng probably sums up the legacy of *Futurama* best in this comment: "Like all great science fiction, it took meaningful messages about modern life and wrapped them up in a package of robots, spaceships, and telepathic toads."[5] Cheng is completely right. Science fiction often puts our society in metaphor to show the things that are wrong and that should be fixed, and the things that are right that should be celebrated. When you live inside of a particular city or even of a particular country, it's hard to look at things with an outsider's fresh eyes and to appreciate the good, the bad, and the improving. But in a science fiction world, we instantly can see things differently. The best science fiction has tackled the issues of gender and equality and freedom and the other things that bother us today—and *Futurama* is no exception.

Cheng has several examples of what *Futurama* taught, but the one that is perhaps most relevant for this argument is to "go against your programming." Cheng cites the pilot episode, where Bender (a robot) and Fry are stuck inside a closet inside the New New York Head Museum. As Cheng explains:

It is in this pilot episode, several fans (including Cheng) argue, that Bender becomes more than just a robot—rather, becomes a being that is flexible to the circumstances. In fact, that's a life lesson right there—that it's so important to be adaptable and ready to alter in the face of changes in your family, your job, or other aspects of your life.

Futurama has also had to be adaptive with the circumstances. It initially ran on Fox, ceased its run in 2003, and then rebooted on Comedy Central in 2008 for a last five-year run. For a series to disappear off the air and then return shows quite the staying

5   Cheng, Gabriele. (2018, Aug. 15). "What I Learned from Futurama." Pop Cultural Studies. Retrieved from https://popculturalstudies.wordpress.com/2018/08/15/what-i-learned-from-futurama/

power with fans (and I'm sure that it's something that devotees of *Firefly* envy of *Futurama*, but that's another story).

Does *Futurama* still have relevance today in the years after its cancellation? One can argue that the series was in fact ahead of its time, so it may become something that future generations appreciate because it looked to something that few of us thought about ten years ago. Take the politics of gender, for example—there is an entire scientific paper available in the journal *Organization* arguing that the episode "Raging Bender" "explicitly engages in drag-based gender parody . . . [which] relates to the politics of doing and undoing gender in organizations more generally."[6] I would like to think that we are even more sensitive to gender than society was at the time of this 2000 episode, but many advocates in the transgender communities (and other communities) argue that we have a long way to go. Did *Futurama* go far enough? I'll leave it to these communities to make the decision.

6   Rhodes, Carl and Pullen, Alison. (2012, June 25). Parody, subversion and the politics of gender at work: the case of Futurama's 'Raging Bender'. *Organization*. https://doi .org/10.1177/1350508412447246. Retrieved from https://journals.sagepub.com/doi /abs/10.1177/1350508412447246?journalCode=orga&

DEJA VU
(2006)

How badly could time travel affect our privacy? While the two concepts aren't often linked in the literature, *Deja Vu* argues that yes, they actually are. The characters live with a massive surveillance program that spies on people "real time" in the past, all in the name of solving murders or doing other things that we imagine are good. While *Deja Vu* doesn't wrestle much with the ethics of this situation, we certainly can in this chapter.

The Science of Time Travel

Characters in *Deja Vu* explain time travel so badly that in frustration, Douglas Carlin (played by Denzel Washington) dramatically picks up a ginormous old-school cathode-ray-tube computer monitor. "Here, look. Here's a monitor, right?" He throws it aside and the monitor breaks apart. "Now the monitor is broken. It's dead. It is not 'temporarily transitioned to another state of entropy.' It's dead, right?" Let's try and explain what's happening, a little bit more slowly (lest you smash your nearest device in frustration at the explanation). At first, Carlin is told that what he is watching is an elaborate surveillance program called Snow White, which plays back footage taken four days and six hours ago. The program has so much data being thrown at investigators that nobody can archive all the footage. All you can do is watch one stream at a time and record that single stream. But Washington begins to see things that cannot be explained by footage alone, as it appears Claire Kuchever (Paula Patton) is reacting to the camera view as it "hovers" around her. So finally, we get the real story—it's not a surveillance project per se, but a time-machine project that uses surveillance.

As characters explain in the movie, here's the scoop. Cambridge University did research and development under a National Reconnaissance Agency grant. The stated purpose was to make telescopes more sensitive when they are looking at the stars. (Astronomers *love* this kind of sensitivity, because if they can see more stars then they can draw more conclusions about how stars are "born" and "die," or how they interact with each other.) Cambridge tried to improve telescope technology by using energy bursts. (It's not made clear how the energy is supposed to make telescopes see better, but let's just go with the flow of the plot.) Turns out their project was generating so much energy that not only did they cause a huge power outage in the United States and Canada,[1] they generated a time warp. How? By creating something called an Einstein-Rosen bridge—a fancy name for a wormhole. Let's go a little deeper into how this "bridge" works. We'll have to take a brief journey into theoretical physics to understand why, but we promise to keep it short and to the point. Physicist Albert Einstein explained that any mass "warps" space and time around it. To take a simple example from How Stuff Works, imagine that two people are holding a bedsheet tight between them. One of those people (with their spare hand) places a baseball on the sheet. The baseball naturally rolls to the middle of the sheet, and the sheet curves around it.[2] We can see this phenomenon working in real life as well. Remember Chelyabinsk, the asteroid that broke up over Russia in 2013? The asteroid fell into the "curve" that Earth's mass produces in space. Or, to put it in terms we use every day, the Earth's gravity trapped the asteroid and it fell into our atmosphere.

---

1   The massive power shortage, which affected millions, actually did happen on the Eastern Seaboard in August 2003, but as far as we know it's not due to time travel.
2   Lamb, Robert & Bonsar, Kevin. (2019). "How Time Travel Works". How Stuff Works. Retrieved from https://science.howstuffworks.com/science-vs-myth/everyday-myths/time-travel4.htm

Okay, so what does this have to do with time travel? Think back to that baseball curving the space around it. Now imagine that there was a second sheet with a little marble. An Einstein-Rosen bridge between these locations would allow people to instantaneously transport themselves. (How this bridge actually arises between the locations takes some fancy physics to explain, but as we did with the plot of *Deja Vu*, let's just assume it could happen given the right conditions.) Or, as How Stuff Works explains[3] it:

> In space, masses that place pressure on different parts of the universe could combine eventually to create a kind of tunnel. This tunnel would, in theory, join two separate times and allow passage between them. Of course, it's also possible that some unforeseen physical or quantum property prevents such a wormhole from occurring. And even if they do exist, they may be incredibly unstable.

Deja Vu (rightly) points out that humans can't see this theoretical bridge, but it's just as "real" as a cell phone signal or radio wave. But here's why Carlin smashes the computer monitor apart. It's really difficult to explain if Kuchever is alive or dead. In the present, she's certainly dead—her body was at the morgue and we saw Carlin examining it. But in the past, she is reacting to the surveillance in real time—so she is alive. So, can we warn her? In the "universe" of *Deja Vu*, we find out it is possible to send living things back through time across this bridge, but the issue is that all electromagnetic activity in the body stops. Why is that a problem? Your heart stops beating, for one. The clever solution somebody proposes is to send back a person to an emergency room, so that their heart can be restarted by a physician.

---

3   *Ibid.*

We've probably already gone a bit too spoiler-y in this description, but hopefully this helps you better understand the concept of a wormhole. We should emphasize here that although Einstein's and Rosen's physics support the idea of wormholes, they remain highly controversial—and may not exist in real life. Scientists wrestle with the geometry of these wormholes, which apparently can only be stabilized by something called "negative energy density." We know what energy is, but creating a negative energy is something difficult to imagine (and also, impossible for matter to do, as *Scientific American* points out). Noted physicist Stephen Hawking even went so far as to say wormholes are impossible to use for time travel because even one particle inserted into the tunnel would destabilize the bridge.[4] "Wormholes are great theoretical fun, and are seemingly valid solutions of the Einstein equations. There is, however, no experimental evidence for them," Richard F. Holman, a Carnegie Mellon University physics professor, said in the article. But he encourages other uses of wormholes: "This should not stop any budding science-fiction writers from using them as needed."

The Science of Privacy

Surveillance is a fairly recent innovation in films. Although noir detective stories in the 1940s dealt with watching people, the watching was used by "a guy in a fedora hat, standing in the shadows, smoking cigarettes while keeping an eye on a building across the street," writes Nicole Rafter.[5] One of the first major films to address surveillance was *The Conversation* (1974), where the main character, Harry, uses an old-school tape recorder to follow the

---

4   Holman, Richard, Hiscock, William and Visser, Matt. (n.d.) "FOLLOW-UP: What exactly is a 'wormhole'? Have wormholes been proven to exist or are they still theoretical?" *Scientific American*. Retrieved from https://www.scientificamerican.com/article /follow-up-what-exactly-is/

5   Rafter, Nicole. (2007, Feb. 22.) "Surveillance and Spying in Film: I-Deja Vu." *Oxford University Press Blog*. Retrieved from https://blog.oup.com/2007/02/surveillance_an2/

conversation of a man and woman in a park—a conversation that quickly leads Harry to question his own sanity. Rafter says that *Deja Vu* didn't push the concept further, even though the technology was more futuristic: it is "window dressing for an old-fashioned detective tale," she says, with the surveillance used "to jazz up an otherwise dull cop film rather than to make a statement about the condition of contemporary society."

While humans have been using technology ever since they had tools, it's only in recent years that we've really worried about privacy. I hesitate to write here about specific technologies, because the landscape evolves so quickly that what I say could be out of date within a year or two. But there are few general examples that are true across *all* technologies.

Sharing information on social media connects communities, but it can also be exploited by trolls, advertisers, or people who are more interested in stalking you than benefiting from your knowledge. Internet searches (no matter what device we use) help us gain knowledge, but companies track these searches through cookies and similar technology to sell that information to advertisers. Even going outside of the house and leaving behind your personal device leaves you no guarantee of privacy. Britain has been criticized for being the most surveilled society in the world, as cameras and other devices (linked in a central network) keep track of citizens on the street, all in the name of protection. In late 2018, the European Court of Human Rights ruled that the United Kingdom's mass surveillance program violates human rights.[6]

Deja Vu takes that surveillance concept and makes it even more creepy, including sound and multiple camera views that you can manipulate in "real time." And to the analysts' credit, not everyone

---

6   Agerholm, Harriet. (2018, Sept. 13). "UK Mass Surveillance Programme Violates Human Rights, European Court Rules. *The Independent*. Retrieved from https://www .independent.co.uk/news/uk/politics/uk-mass-surveillance-gchq-eu-human-rights -echr-edward-snowden-a8535571.html

is comfortable with the constant surveillance. During one scene in *Deja Vu*, the camera view lingers on a view of Kuchever scrubbing herself in the bathroom. A female analyst asks, "Is there any scientific or forensic insight likely to be gained by spying on this woman in the shower?" Instead of taking her concern seriously, another male character (among a team who seemed to be enjoying the view) jokes: "We're trying to make sure the woman's clean."

So, what is the solution? It seems that every new Internet-connected technology is a bear's trap for anyone looking to stay discreet. "Privacy is the elephant in the room, unavoidable at every turn, and too big to ignore," wrote a CBC News columnist in 2017, "and it will keep coming up week after week, especially as the Internet of Things evolves from being a concept to a new reality. As the world's biggest tech companies all race toward the next big thing, maintaining our privacy seems at risk of being bulldozed."[7]

The first challenge is understanding that technology will *always* be an arms race. For every time that somebody makes a hard-to-break code, there's a hacker trying to break it; even though it's sometimes for good reasons (like building better systems), it makes for a discouraging proposition to declare that your information is secure. There seem to be three options, as advocates pose it—not to share information in the first place, to call for more regulation for how information is stored and shared, or to use some defensive technology to promise us security.[8] It's too much of a Gordian knot for any society to unravel quickly, but it's helpful to understand the shape of the debate for consumers to make informed decisions.

---

7   Pringle, Ramona. (2017, May 2). "In the World of Technology, It All Keeps Coming Back To Privacy." *CBC News*. Retrieved from https://www.cbc.ca/news/opinion/tech-privacy-1.4093996.

8   Swanson, Bret. (2018, Sept. 27). "The Best Solution To Digital Privacy Challenges? More Technology, Not European-Style Regulation." *American Enterprise Institute*. Retrieved from http://www.aei.org/publication/the-best-solution-to-digital-privacy-challenges-more-technology-not-european-style-regulation/

The Legacy of *Deja Vu*

While *Deja Vu* is more spy thriller than a commentary on the impact of technology on privacy, at least one critic feels that the movie helps viewers process tragic events of the past. "There are visual reminders of the Katrina catastrophe, and even references to the Oklahoma City bombing, which happened way back in 1995," writes Peter Bradshaw in the *Guardian*.[9] As the National Weather Service describes it, Katrina was a "large and extremely powerful hurricane" that killed 1,833 people and cost about $108 billion in 2005 dollars.[10] Among the hard-hit cities was New Orleans, the backdrop for *Deja Vu*—which was released only a year after the natural disaster. As for Oklahoma City, this is what happened: Two coconspirators with anti-government sentiment set off a truck filled with explosives outside the Alfred P. Murrah Federal Building in Oklahoma City. The resulting blast killed 168 people and was the deadliest terrorist attack in the United States until the New York City and Washington, DC terrorist plane crashes of Sept. 11, 2001.[11]

Why does the discussion of history have so much relevance in *Deja Vu*? As we have seen, privacy is not a problem unique to our decade—it's something that has been explored in film since the 1970s (at the least). And in the wake of 9/11, one can argue that authorities are monitoring our communications more in the name of security. American academics Jack Balkin and Sandy Levinson once called the United States a "national surveillance state" because (in their view) there are so many departments and bureaucratic institutions dedicated to collecting digital information with little

9   Bradshaw, Peter. (2006, Dec. 15). Denzel Washington: Deja Vu. *The Guardian*. Retrieved from https://www.theguardian.com/film/2006/dec/15/denzelwashington.actionandadventure

10 National Weather Service. (August 2005). "Extremely Powerful Hurricane Katrina Leaves a Historic Mark on the Northern Gulf Coast." Retrieved from https://www.weather.gov/mob/katrina

11 History.com. (2009, Dec. 16). Oklahoma City bombing. Retrieved from https://www.history.com/topics/1990s/oklahoma-city-bombing

scrutiny, especially after 9/11 occurred.[12] What *Deja Vu* exposes is that fine line between conducting surveillance for security, and conducting surveillance in a way that is intrusive and hurts a person's privacy. Here's the real trouble—that "line" varies for each one of us. What one person considers a safe practice may make another person feel deeply uncomfortable; look at the number of people who blithely install apps that pull personal information from your phone such as contacts or text messages. Is this something that app needs? How comfortable are you with a tech company possessing this information? Unfortunately, in our hurried world it is difficult to answer these questions when you need that app rapidly to get work done.

What's worrisome is that today, there are so many devices competing for our attention that it is easy to get distracted. While we power through shows on Netflix, the streaming service collects information about our browsing habits to better feed our interests. While we surf the Internet on our devices, companies follow our digital footprints to send ads and other information our way. In many cases, what we see presented to us on our screens is what we *want* to see—a collection of search results and advertisements personalized to our browsing habits. So, in this age of shrinking privacy, it may be our ability to read critically and think about the consequences that comes most under threat.

---

12 American Bar Association. (2017, June 30). "Privacy and Surveillance Post-9/11." Retrieved from https://www.americanbar.org/groups/crsj/publications/human _rights_magazine_home/human_rights_vol38_2011/human_rights_winter2011 /privacy_and_surveillance_post_9-11/

MEET THE ROBINSONS (2007)

This sweet Disney film should get a lot more attention. It's an uplifting story about how a kid with fewer advantages than most—living without a stable family and struggling to find his own identity—learns a little more about the value of trial-and-error from a mysterious visitor from the future. I really found myself rooting for Lewis and enjoying his maturing as the movie went on, and I bet you will too. My only complaint is that this movie leaves a lot of the "time-travel" science up to interpretation.

The Science of Time Travel

In *Meet The Robinsons*, the first person we meet and get to know is Lewis. Lewis is twelve years old and really works hard at building machines. But he struggles in other ways—he lives in an orphanage and is haunted by the memory of his mother abandoning him there when he was a baby. Lewis creates a memory scanner and brings it to his school's science fair, where he runs into thirteen-year-old Wilbur Robinson, who says he's a "time cop" from the future. After a few mishaps and a run-in with a "bad guy," the two boys suddenly find themselves, indeed, in the future. Lewis gets to meet the Robinson family and to learn a little more about how to feel better about himself. There's a haunting plot twist near the end, and then Lewis needs to take his "learning" to help others. Not a bad moral message, at the end of the day.

But yeah, this is supposed to be a book about the science of time travel, so let's zoom in on how Lewis and Wilbur get to the future. Predictably, it's by flying car. I mean, this time-travel machine is from the future, so we have to have flying cars, right? But remember, this is a Disney film, so we don't get a lot of detail on how this time machine works. We learn that one of the Robinsons worked hard to invent this machine, through a lot of trial and error, and that when you climb in this car you can alter "the time stream." But at the end of the day, all we know about this machine is that even a couple of kids can climb inside, press some buttons, and zoom into the future—or the past.

In trying to figure out how the kids did it, I took a careful look at the animation during the first trip that Lewis and Robinson take together on screen. The car zooms up into the sky, gets surrounded by a bubble, and then appears to go into a "time tunnel" (that's my term, not theirs) before popping out in the year 2037. So I was wondering—is the bubble key to time travel itself? A 2017 paper led by a University of British Columbia researcher[1] does suggest that "bubbles" could make time travel possible into the past. Ben Tippett borrowed from the equations of Einstein and general relativity to create a theoretical time machine, which a UBC press release describes thus: "A bubble of space-time geometry which carries its contents backward and forward through space and time as it tours a large circular path. The bubble moves through space-time at speeds greater than the speed of light at times, allowing it to move backward in time."[2] Tippett explained a phenomenon of Einstein's space-time model, which explains that in the presence of a huge mass in space, like a black hole, anything caught in the gravitational well would experience time dilation. Time would move slower. Tippett's bubble "time machine" exploits a curve in space time to "bend time into a circle for the passengers, not a straight line," as he explained it. "That circle takes us back in

---

1   Tippet, Benjamin K. and Tsang, David. (2017, March 31). "Traversable Acausal Retrograde Domains in Spacetime." Classical and Quantum Gravity 34(9). Retrieved from https://iopscience.iop.org/article/10.1088/1361–6382/aa6549/meta#cqgaa6549s2

2   Wellborn, Patty. (2017, April 27). "UBC Instructor Uses Math To Investigate Possibility Of Time Travel." The University of British Columbia Okanagan Campus. Retrieved from https://news.ok.ubc.ca/2017/04/27/ubc-instructor-uses-math-to-investigate-possibility-of-time-travel/

time." But he warns that this theoretical time-bubble cannot exist in real life, because physicists need to use something called "exotic matter" to allow it to work.[3]

So what is exotic matter? A BBC website explains that it's basically anything that is different from normal matter, such as "superfluids" that sound like the plot of many horror films: "isotopes of helium whose quantum properties allow them to defy gravity, escaping from containers by creeping up and over the walls."[4] (Eeeeeew.) Let's abandon this remake of *The Blob* and instead focus on a form of exotic matter that most of us are a little more familiar with: dark matter.

Dark matter has its roots (if we can call it that) in a 1998 Hubble Space Telescope discovery that the expansion of the universe is not slowing down—in fact, it's accelerating. While scientists squabble over what exactly is happening, we'll focus on one explanation: dark matter and dark energy. "Normal" matter is the stuff of the universe that we can see: stars, galaxies, planets, you name it. "Dark matter" is not visible using our conventional telescope instruments, and scientists aren't sure what kind of particle this dark matter would be made of. NASA explains that it does seem to exist, because it produces an effect called "gravitational lensing" that allows large masses in space (such as galaxies) to focus the light from more distant objects behind them, making them visible in telescopes. Meanwhile, "dark energy" affects the expansion of the universe; perhaps that energy is a property of the universe itself, but we poorly understand why.[5]

3 *Ibid.*
4 Matthews, Robert. (n.d.) "What Is Exotic Matter?" Science Focus: The Online Home of BBC Focus Magazine. Retrieved from https://www.sciencefocus.com/space/what -is-exotic-matter/
5 Nagaraja, Mamta Patel. (2019, Feb. 13). "Dark Energy, Dark Matter." NASA. Retrieved from https://science.nasa.gov/astrophysics/focus-areas/what-is-dark-energy

So, what does this have to do with time travel? Well, if we're going to be able to manipulate exotic particles, first of all we have to know what these particles are. Then we have to know how to find them. Then we have to know if and how to manipulate them to work in our time machine. That's a lot of big gaps of knowledge we have to master first. But hey, science is about learning and optimism that our knowledge will accumulate over time, right? Maybe we'll figure out the solution through hard work. Or at the least, we'll take the learning we have and make even more accurate science-fiction stories over time.

The Science of Memory

The cutest invention of the movie is (surprisingly) not the time traveling machine, but the memory scanner. As Lewis explains it, "First, you input the desired period of time on this keypad. Then a laser scans the cerebral cortex, where memories are stored. The retrieved memory is then displayed on this monitor." Before analyzing this, we need to know, how are memories formed? Why is it so easy to forget that person's name who you just met, but so hard to burn that annoying song lyric from your brain forever? ("Never gonna give you up, never gonna let you down . . . ") In the *Guardian*, neuroscientist Dean Burnett explains memory like this. We use short-term memory much like random access memory on a computer—it's a piece of information (like a phone number) that we hold temporarily in our head to get a task done. We likely conduct this thinking in the prefrontal cortex, which is in the front of your brain.[6] Then there's long-term memory, which is formed when neurons in the brain make new connections and synapses with one another. There are many categories of long-term memory.

---

6   Burnett, Dean. (2015, Sept. 16). "What Happens In Your Brain When You Make A Memory?" *The Guardian*. Retrieved from https://www.theguardian.com/education/2015/sep/16/what-happens-in-your-brain-when-you-make-a-memory

Driving a car? Implicit memory (the automatic habits or skills we memorize). Knowing that Washington, DC is the capital of the United States? Explicit memory of the semantic sort, the kind that falls into "general knowledge." And for those memories of events and milestones in our lives, this is another kind of explicit memory: episodic memory.[7] When an event happens, the brain's hippocampus and frontal cortex work together to process all the inputs coming at you: sight, sound, hearing, taste, you name it. Then they encode this stuff in your brain by connecting the brain cells, using electricity and chemical messengers (neurotransmitters). The big mystery in science today is how we retrieve these memories later on,[8] so it sounds like Lewis is on to something big with his scanner. Maybe his visit to the future taught him something new about the brain.

Meanwhile, we *do* know of tricks to make it easier to recall memories, especially when you're trying to memorize dates or jumbles of facts for tests. There are all sorts of ways to help you out, including associating the facts with people or objects you are familiar with, creating acronyms, testing yourself and writing out the answers, and making sure to get enough sleep.[9] That's right. All-nighters don't allow your brain to rest and retain memory, so don't leave your studying until the last minute.

The Legacy of *Meet the Robinsons*

While *Meet the Robinsons* appears to be a standalone film in terms of its pop culture impact, it is a cute ride for kids—although adults may cringe at the strange interlude in the middle of the film where

7   *Ibid.*
8   Mohs, Richard C. (n.d.) "How Human Memory Works." How Stuff Works. Retrieved from https://science.howstuffworks.com/life/inside-the-mind/human-brain/human-memory1.htm
9   Lickerman, Alex. (2009, Nov. 16). "Eight Ways to Remember Anything." *Psychology Today*. Retrieved from https://www.psychologytoday.com/ca/blog/happiness-in-world/200911/eight-ways-remember-anything

Lewis hangs out with his adopted family from the future. There's a lot of incidental and slapstick humor in there, but not much material to move the plot forward. Instead, I'd like to pick up this theme from the blog *Pop Culture Pundit* which reviewed the film a few years ago. Author Steven Johnson said: "There are two key themes running throughout this film: first, the importance of a space where creative minds can tinker, with room for trial and error. Second, the reality that you need to work to change your own future: no one is going to do that for you."[10] Or as the line from *Meet the Robinsons* goes: "From failing, you learn. From success, not so much. If I gave up every time I failed, I never would have made the meatball cannon. I never would have made my fireproof pants." (Side note: Can I have a pair of fireproof pants, please? It'll help me next time I turn on the stove.)

Let's go with Johnson's second point first, that your future is not fated and that you have a role in changing it. Fate is tied into the philosophical definition of fatalism, which (simply put) says that we aren't in control of our actions and simply act by predetermined conditions. Some people cite God in this argument, while others go a little more scientific and discuss causes and effects.[11] This isn't a book about the science of philosophy, so I won't get into the arguments and discussions about fatalism from noted thinkers such as Aristotle, Richard Taylor, or Nelson Pike. Except to say, this idea of fatalism is *super* depressing. Why bother study for that test, if we're fated to fail it anyway? Believe me, it's not a fun way of living. Perhaps a more healthy approach to living would be to think like a Stoic, classifying events in life over which you have

---

10 Johnson, Steven. (2016, Feb. 19). "Meet The Robinsons (2007)". *Pop Culture Pundit*. Retrieved from https://popculturepundit.wordpress.com/2016/02/19/meet-the-robinsons-2007/

11 Stanford Encyclopedia of Philosophy. (Written 2002, Dec. 18; revised 2018, Dec. 5). "Fatalism." Retrieved from https://plato.stanford.edu/entries/fatalism/

some control, no control, or complete control.[12] I'd like to hope passing a test is at least partially within my control, even though I can't stop a snowstorm from descending on me on my way to the exam. (I live in Canada, after all.)

Now for the first part of Johnson's argument, about the importance of tinkering, and trial and error. I love how the film shows Lewis trying and failing again and again to build a machine, or meeting the right people who will be his parents. Struggling in life is just a fact of life because we all have our strengths and weaknesses. Maybe you sail through algebra in math class, but calculus hurts your brain. Or you're a wizard at deconstructing Shakespeare, but in your school's required foreign-language class you struggle to put together a sentence. How do we get through challenging circumstances? There are at least a few approaches. Perhaps the easiest one (in terms of immediate effort) is just giving up, which Lewis vows to do early in the film. But as he meets people outside of his immediate circle of friends and adopted family, he learns that we all make mistakes and there is more than one way of solving the problem. Perhaps a tutor will help you crack calculus, or some online videos on Khan Academy. Or perhaps it's just time and effort and trusting the process of learning.

12 Irvine, William B. (2008). "A Guide to the Good Life: The Ancient Art of Stoic Joy." Oxford: Oxford University Press.

11/22/63
(2011)

What happens when falling in love gets in the way of you accomplishing something great? What if that "something" was to change the very course of history, which is stopping the assassination of US President John F. Kennedy on Nov. 22, 1963? That's the question that Stephen King tries to answer in his book *11/22/63*. It was an era when the charismatic young president seemed to represent all things future. The romance of the Kennedy era evaporated when the president was gunned down during a motorcade. But was it real romance, or just nostalgia for another time?

The Science of Time Travel

The story gets going when adult high school teacher Jacob "Jake" Epping leaves the present day and enters a sort of time portal to the year 1958 to rescue JFK. He takes on a persona, George Amberson, that he carries through most of this time in the past. (So to make things easier, we will refer to our narrator as "George.") George has five long years to stop JFK's assassination, and as readers will see from reading the book, it's long enough for a man to fall in love with somebody and question whether the sacrifices are ultimately worth the pain.

We don't hear much about the science of George's time-travel portal, or "fissure," as the characters call it. We do know it's at the bottom of some basement stairs in a friend's store. His friend confidently says that every time you go into the portal, time resets—although George later learns that even innocent rambles around the past could leave a residue in the minds of some residents. No matter how long you spend in the past, only two minutes go by in the present when you return. This means you could come back from the past with gray hair, or with a chronic illness, and *really* confuse your friends.

How exactly this portal is constructed is up to your imagination, and luckily we've already gone through some of the main concepts in the book. It could be a wormhole, or a sort of "bridge" in time between two points. As you'll remember, wormholes are more popularly used for travel between two different locations, but in *theory* you could possibly use it to breach time. It's more difficult to imagine the physics where you jump in both time and space simultaneously, and King is careful to follow the rules here; when George emerges in 1958, it's in the same location as the present day.

One thing that King repeatedly emphasizes in his book is the "butterfly effect," a term first coined by Edward Lorenz in a 1963 paper. Simply put, when looking at larger weather systems, Lorenz suggested that even the smallest effect, such as a butterfly flapping its wings, may alter the world so much that a tornado could be generated. This "sensitive dependence on initial conditions," as Lorenz called it, also has an effect in time: it means that forecasting could be a fool's game, something nearly impossible to accomplish.[1] This also means that George, much as he tries to do good in stopping the JFK assassination, could cause some unintended consequences. The simple act of George falling in love with a person from the past also makes some interesting fiction, and time paradoxes. You'll see what I mean when you read this gripping, nearly 850-page book in hardcover. Because I'd hate to ruin the surprise.

The Science of Coincidence

King's horror-writing influence shines through often in this book, especially when talking about the "science of coincidence." Let's

---

1 Dizikes, Peter. (2011, Feb. 22). When the Butterfly Effect Took Flight. *MIT Technology Review*. Retrieved from https://www.technologyreview.com/s/422809/when-the-butterfly-effect-took-flight/

take an example. Early in the book, George wants to do a terrible, bloody deed that he is convinced will change the future of one of his students. But, as we hear over and over again throughout the book, the past is "obdurate," meaning that it doesn't like to change. So George develops the worst of stomach flus that cripples him simultaneously with diarrhea and vomiting. When he staggers into a pharmacy, he's in so much unbelievable pain that the person ringing up his purchases actually laughs at him. George hops up on meds and approaches the house of his intended target, then trips, falling spectacularly in one of the yards. Minutes later, another person stops George in a long, pointless conversation whose ultimate aim is to stop him from doing the hit. It's horror-comedy and we can all relate to this. Just think about those days when you're trying to get to an important meeting. Somehow, spilled coffee, a sick dog, and an unexpected traffic jam all convened upon you in such a way that you were sure The Fates were out to get you. How coincidental are all these coincidences? Surprise, surprise—maybe not so coincidental.

As numerous court examples will show you, eyewitness testimony is a terrible way of seeking accuracy.[2] This may be especially true when you're overwhelmed with emotion (whether good emotions or bad emotions). In thinking about this, I think about the time when I met my husband online. I remember vaguely that we connected to each other only within days of signing up for the service. Coincidental? My best recollection is yes, it was a coincidence in a short period of time. Then again—was this within days of signing up, or within weeks? I can't remember exactly. My mind was so overwhelmed by emotions that I didn't take careful note of my surroundings. Same goes for poor George in *11/22/63*. He

---

2   Hurley, Greg. (2017). The Trouble with Eyewitness Identification Testimony in Criminal Cases. National Center for State Courts. Retrieved from https://www.ncsc.org/sitecore/content/microsites/trends/home/Monthly-Trends-Articles/2017/The-Trouble-with-Eyewitness-Identification-Testimony-in-Criminal-Cases.aspx

may have tripped and run into that bore of a secondary character simply because he was too ill to pay attention to his surroundings.

Even with (impossibly) perfect memory, however, the science of probability works against us. Probability is the science of how often things happen. It's easy (or easier) to calculate probability when rolling a dice, for example, but in the universe things get a lot messier. Look at the number of people who believe the "Face on Mars" (captured in a NASA Viking spacecraft photo in the 1970s) is a real face crafted by aliens to signal something important. It didn't matter how many later pictures NASA released showing that this "face" was a coincidental convergence of a primitive camera and a few shadows.[3] We also can't forget that our brains like familiarity (because patterns make it easier to make decisions) and we'll easily invent faces because they're a familiar and important pattern.[4] Our world in fact has "faces" everywhere if you look for them. The knobs on kitchen cabinets or the headlights on a car can look like eyes.

Some noted scientists have attributed coincidences to something less scientific, such as Carl Jung, who argued for an underlying synchronicity of our world that could also be traced to telepathy, ghosts, and mind-reading (extrasensory perception).[5] But probability destroys a lot of coincidence if you think about the science. A 2017 *Forbes* column by Chad Orzel summed it up beautifully in a headline: "Probability is more certain than you think."[6] From the

---

3   National Aeronautics and Space Administration. (2001, May 23). Unmasking the Face on Mars. NASA Science. Retrieved from https://science.nasa.gov/science-news/science-at-nasa/2001/ast24may_1

4   Martinez-Conde, Susana. (2018, May 1). The Fascinating Science Behind Why We See 'Faces' In Objects. Mental Floss. Retrieved from http://mentalfloss.com/article/538524/science-behind-pareidolia

5   Beck, Julie. (2016, Feb. 23). Coincidences and the Meaning of Life. *The Atlantic*. Retrieved from https://www.theatlantic.com/science/archive/2016/02/the-true-meaning-of-coincidences/463164/

6   Orzel, Chad. (2017, May 3). In Science, Probability Is More Certain Than You Think. *Forbes*. Retrieved from https://www.forbes.com/sites/chadorzel/2017/05/03/in-science-probability-is-more-certain-than-you-think/

vagaries of US politics to the weird way that subatomic particles interact in quantum physics, Orzel argued, probability makes strange things align from time to time. After all, if the Large Hadron Collider runs hundreds of millions of particle collisions every second, you're bound to see some unlikely combinations crop up again and again. "There are no shruggies in science," Orzel writes, and that makes me take the face value of coincidence in King's work a little less seriously.[7]

The Legacy of *11/22/63*

I admit I came down a little hard on King's work at times. So, let's be clear about this—despite some of the scientific flaws, his book is my favorite time-travel story of all time. The book is about dates with destiny (yes, including literal dates with romantic partners) and how this affected a crucial period of American history before and after Kennedy was assassinated. The Kennedy assassination is such a pivotal cultural event that it is difficult to imagine a world where he had somehow survived. As the decades mounted and the pain waned, American mass media began to look sardonically at the event—such as a famous Seinfeld two-part episode involving a spitting baseball player, or the character The Comedian from *The Watchman* among other examples.[8] In other words, the cultural event of the assassination spawned other references that in themselves became cultural events. Today is a much different world because of Kennedy's death. As one columnist wrote about King's book in 2011, "Stephen King in *11/22/63* confirms that our world is not a stage for pop culture but a total environment

---

7   *Ibid.*

8   Zimmerman, Joel. (2013, Nov. 16). The Top 7 JFK Assassination Pop Culture References. Rewire. Retrieved from https://www.rewire.org/pbs/top-7-jfk-assassination -pop-culture-references/.

constructed by pop culture, an environment in which this novel, without illuminating the surround, fits right in."[9]

Would the United States have been a better place if Kennedy escaped unscathed? This book honestly answers the question, and you might be surprised at its conclusions. For me, the takeaway from the book was learning to live life and accept the past for what it is. We all make mistakes in our personal histories. That proof you shouldn't have ignored, that person you shouldn't have dated. We get emotionally scarred and tell ourselves that it was really all a stupid mistake. But the real value of these experiences, if we can call it a value when reflecting upon horrible past events, is learning from them and then trying to use that knowledge to help others.

Stories and storytelling have been part of everyone's cultural history long before the Mahabharata, or Homer's *Odyssey*, or whatever equivalent you prefer to use from your particular culture or region of the world. We learn from the experiences of others. What experiences do you have that will help the younger set succeed? The more you think about that question, the better this world is going to be. So reflect on that carefully.

9   Marchand, Philip. (2011, Nov. 11). Open Book: 11/22/63, by Stephen King. *The National Post*. Retrieved from https://nationalpost.com/afterword/open-book-112263-by-stephen-king

LOOPER
(2012)

As an experienced time-travel veteran, you now know about stuff such as the butterfly effect or the grandfather paradox. What's fun about *Looper* is it plays with both of these concepts in an unexpected way. The movie's simple premise is that it sets up situations where young people kill off their older selves (who travel back in time to meet their younger selves) for a stack of riches. But what if that older self escapes the assassination and causes havoc in the past? How does that affect the past—and future?

The Science of Time Travel

Looper not only does a poor job of explaining time travel, but the screenwriters revel in it. One of the funniest lines of this otherwise sad movie comes when old Joe (Bruce Willis) and young Joe (Joseph Gordon-Levitt) begin a discussion about the mechanics of time travel. In typical senior-citizen style, old Joe just skips over the conversation because he feels it's wasting time. "I don't want to talk about time travel because if we start talking about it then we're going to be here all day talking about it, making diagrams with straws," he snipes. But here is the little we can know about time travel, directly from the plot. It's the year 2044 and young Joe is a "looper" working for a crime syndicate. His job has to do with time travel; thirty years in the future, the syndicate creates time travel because future technology systems make it so hard to dump bodies after you do a Mob-style "hit." The syndicate's unique solution is to take their intended victims and force them in a time-travel machine. They send the victim back to a specified time and place in the past and a "looper" such as young Joe hangs around the location, shotgun in hand, to kill the masked target.

For every "kill" a looper racks up, they receive a bunch of silver bars. Young Joe is actually more forward-thinking than most loopers, as he saves a good deal of these bars in a secret compartment in his apartment. But here's the rub—no matter how responsible you are as a looper, your very job signs your death warrant. The last "kill" you make will be your own self in the future. You'll know it's your self because that last victim will have gold bars strapped to them. "This is called 'closing your loop,'" young Joe explains. "You get a golden payday, a handshake, and you get released from your contract. Enjoy the next thirty years. This job doesn't tend to attract the most forward-thinking people."

So, how does this time machine work? Doesn't it create some really weird paradoxes? Without getting too heavy into the spoilers, yes, you can imagine that when young Joe encounters old Joe and they begin to work toward their separate goals, there are issues in terms of timeline consistency. We like to imagine that our lives move forward in an irrevocable timeline, that the actions we take in the past affect our present and future. To think of present and future actions affecting our past scrambles our minds; it's very hard for us to imagine. But let's try and stay with it for a moment.

Looper explores (in some details) a phenomenon known as the "causal loop." As the *Atlantic* explains it, a causal loop is a large circle or cycle of events. Each event causes the other, and the loop has no beginning or end.[1] So in the movie, the older Joe goes back in time to speak with the younger Joe. Because of older Joe's conversation, younger Joe changes some decisions in his future. Then in thirty years, younger Joe (who is now old Joe) finds himself

---

1 Fetters, Ashley. (2012, Oct. 3). "Time Travel in 'Looper': Dubious, but Not for the Reason You Might Think." *The Atlantic*. Retrieved from https://www.theatlantic.com/entertainment/archive/2012/10/time-travel-in-looper-dubious-but-not-for-the-reason-you-might-think/263192/

in the time machine and speaks with the next version of younger Joe, who starts the cycle anew.

It's complicated for me to explain here, and likely harder for a time traveler to execute. The *Atlantic* points out that some theorists (such as New York University analytic philosopher Paul Horwich) think this chain of events is too improbable.[2] Just think about it, real life is too messy. Sometimes you're at a grocery store and just happen to spot a friend in the same aisle. What are the chances that the two of you will be there at the same time? While we have argued before that coincidences in life are bound to happen *in general*, the odds of a specific coincidence lining up is pretty low. How do we know younger Joe will always react to older Joe in the same way? What if older Joe talks to younger Joe at a slightly different moment or says a slightly different thing? It's too hard to imagine.

That said, not every physicist is skeptical of causal loops. Back in the early 1990s, astrophysicist J. Richard Gott wrote a paper trying to imagine Einstein's equations in a 2D universe, according to *Popular Science*.[3] He found a method of mathematics that does allow for backward time travel, but can we just point out again that this a 2D universe? If one likes breathing, it's probably not a good idea to try this method. Then in 2005, MIT theoretical physicist Edward Farhi (a noted skeptic of time travel—he says that going backward in time would suck most of the energy out of the universe) held what he felt is the ultimate test of time travel. He and his students advertised a time-traveler's convention. They bought milk and cookies. Unfortunately, nobody showed up. Maybe the food selection turned them off? Farhi has another explanation. "Maybe they were clever enough to remain incognito, but none

---

2   *Ibid.*

3   Mosher, Dave. (2012, Sept. 27). 'Looper' And The Real Science of Time Travel. *Popular Science.* https://www.popsci.com/science/article/2012–09/emlooperem -and-real-science-time-travel#page-2

came," Farhi told *Popular Science*. "So, as far as we know, reverse time travel can only exist in the movies."[4]

Other physicists argue that time travel *is* possible, but it would only happen with the smallest particles in the universe—on the quantum scale.[5] Although believe me, studying quantum physics messes with your sense of time in any case. You ever heard of "spooky action at a distance"? It turns out that observing one quantum particle can affect a paired quantum particle—even if the two are separated by light-years. Yup, they appear to be communicating faster than the speed of light, which physicists still struggle to explain.[6] So causal time travel will continue to complicate our equations, whether we talk about the very big or the very small.

The Science of Telekinesis

So this is a super-depressing film, isn't it? Let's focus on one of the more fun parts of it, psychokinesis or telekinesis, or the ability to move things around with the power of your mind. In the "Looper" universe, a telekinesis (TK) mutation spontaneously appears in a small percentage of the general populace. But in most cases, it's more of a party trick. To paraphrase young Joe's explanation of the matter, TK powers amount to a bunch of show-offs "thinking they're blowing your mind by floating quarters." He later discovers somebody who actually uses TK to terrifying effect, especially when that person gets angry. But we'll let you watch the movie to see the result.

4   *Ibid.*

5   Horowitz, Gary T. and Friedman, John L. (n.d.) "According to current physical theory, is it possible for a human being to travel through time?" *Scientific American*. Retrieved from https://www.scientificamerican.com/article/according-to-current-phys/

6   Popkin, Gabriel. (2018, April 25). "Einstein's 'spooky action at a distance' spotted in objects almost big enough to see." *Science*. Retrieved from https://www.sciencemag.org/news/2018/04/einstein-s-spooky-action-distance-spotted-objects-almost-big-enough-see?r3f_986=https://www.google.com/

As *Encyclopedia Britannica* explains, "scientific evidence supporting the existence of psychokinesis is lacking." That's not for lack of trying. Some people have run studies asking their subjects to change the direction of thrown dice, while others asked subjects to change the output of random number generators. The results are . . . controversial. Sometimes the scientists see an effect. Sometimes studies suggest that phenomena such as publication bias (publishing from a set of publications rather than trying to span the literature) or confirmation bias (our tendency as humans to search for information that reinforces our views) influence the findings.[7]

While we can't prove telekinesis happens in real life, it's a fun topic to explore in fiction. Just look at the number of franchises that use it: the Jedi and Sith who use "the Force" in *Star Wars*, Carrie White in Stephen King's novel *Carrie*, and even the character Eleven in the Netflix series *Stranger Things*. And telekinesis is actually a great introduction to something we *can* show in real life—the effects of our minds on our own selves. Scientists need to be adept at avoiding any perception of bias, which is difficult because we are humans. Humans tend to assimilate information rapidly to come to conclusions, which is great when you're escaping a tiger but tends not to be so great when, for example, you're making a decision about what college to attend or whether you should marry the person you're dating.

We've already covered confirmation bias and publication bias here, but there are other cognitive traps to consider. A small but incomplete list includes groupthink (the tendency to agree with the people around you), culture (if you grow up in a Eurocentric culture, you have a different point of view than say, an indigenous culture), the halo effect (deciding that everything about a subject

7   *Encyclopaedia Britannica*. (2019). Psychokinesis. Retrieved from https://www.britannica
   .com/science/psychokinesis

is positive because of a single positive effect) or habituation (answering similarly worded questions to save energy and time). To think critically, we need to be aware of our biases and to slow down and go through the information we have methodically. A perfect example of this comes during election time. Candidates love to one-up each other with one-liners during debates, or to put out attack ads about their opponents, because they assume a certain percentage of the population will *just believe them*. Don't be that sucker. Read the platforms carefully. Study up on each candidate's voting history, or previous employment, to understand their viewpoints. Then make the best decision you can based on the facts—not what the candidate wants you to think, but what you actually think based on what you read and learned.

The Legacy of *Looper*

While on its surface Looper is a time-travel story, there's a more interesting subtext to the plot. We know this subtext is deliberate because both Gordon-Levitt and writer-director Rian Johnson confirmed it in interviews. What is this subtext? Breaking the cycle of violence. Johnson said the movie is his answer to the question about whether (if it is possible) you would travel back in time to kill Hitler, before the outbreak of the Second World War. And Gordon-Levitt argued that "violence begets violence" and that you need to find another solution to stop the awful cycle.[8] This is a question worthy of a few graduate-level theses, but do understand that philosophers through the ages have wrestled with whether violence is justified. Philosopher Avram Alpert argues that Immanuel Kant, for example, was a fan of the ideals of the French Revolution—but not the violence it produced. He says that

8   Kendrick, Ben. (2012, July 14). "Rian Johnson & Joseph Gordon-Levitt on 'Looper' Science: Would You Kill Hitler?" ScreenRant. Retrieved from https://screenrant.com /looper-science-rian-johnson-joseph-gordon-levitt/

Kant, Rousseau, and other philosophical spectators of violence try to focus on the causes of violence, rather than agreeing to "the morality of reactive violence itself."[9] But not all philosophers may be passive spectators, with some people arguing that Aristotle's definition of "virtue" may allow violence to happen (in the ethical sense) as long as the "end," the result, is a good thing.[10]

Does violence beget violence? Is it true that by the old saying, an eye for an eye would leave the entire world blind? In *Looper*, the answer couldn't be clearer. Violence is shown routinely and horrifyingly because it targets the most vulnerable people imaginable—children. Yes, one of these children does grow up to be a violent person, but when that person is young and malleable, is that the time to kill them? Or later on? Or really, ever? Here's a related question, would doing violence against a single person really have that large an effect on the past, present or future? As one blog asked, "Are the fates of people and societies 'overdetermined,' as historians might put it, by structures of power and privilege that make only certain outcomes possible, regardless of what anyone does?"[11] It's not so easy to answer in a few sentences, is it? Western society loves the idea of history being shaped by a single person or a small group of people, but we forget that often society itself has structures in place that favor power to a select group of individuals. We also forget that nameless groups of people can change history forever, such as how the Winnipeg General Strike in 1919 spawned sympathy strikes in other locations and is often cited heavily in

9   Alpert, Avram. (2016). "Philosophy Against and in Praise of Violence: Kant, Thoreau and the Revolutionary Spectator." *Theory, Culture & Society* 33(6), 51–73.

10   Bufacchi, Vittorio. (2013). "Introduction: Philosophy and Violence." In Revue internationale de philosophie 3(265), 233–235. Retrieved from https://www.cairn.info /revue-internationale-de-philosophie-2013-3-page-233.htm#

11   Cummings, Alex Sayf. (2012, Oct. 22). "Time, Fate, and the History of the Future in 'Looper.'" Tropics of Meta. Retrieved from https://tropicsofmeta.com/2012/10/22 /time-fate-and-the-history-of-the-future-in-looper/

histories of labor relations (even though it arguably didn't provide much effect at the time).[12]

The climax of *Looper* shows one solution to end the cycle of violence, and it's not a passive solution. It works within the plot of the film, but is it the way to go about things in real life? I wonder about this every time I watch *Looper*, and the more I read about philosophy and history, the more confused I get. Perhaps that's how adulthood works; the older you get, the less black and white the world appears.

---

12 Reilly, J. Nolan and Skikavich, Julia. (written 2006, Feb. 7/update 2015, March 4). "Winnipeg General Strike". *The Canadian Encyclopedia*. Retrieved from https://www .thecanadianencyclopedia.ca/en/article/winnipeg-general-strike

PREDESTINATION (2014)

While many time-travel movies cover the question of "free will" vs. "fate," if we really do have agency over our future, this one goes at it in a really, really wild way. In *Predestination*, different characters cross paths in a way that we don't expect. And it all starts with a clichéd conversation in a bar, a situation where one character tells the other that they have the craziest story ever to tell.

The Science of Time Travel

Honestly, I found the name of the movie *Predestination* really hard to remember, for the first few weeks after watching it. (I had the same problem with 1991's *Backdraft*, which refers to a fire phenomenon—with family members, I began jokingly calling it "Crankshaft" or "Gearshift" when the name slipped my tongue. Yes, I'm old.) That problem changed when I looked up the actual definition for the word "predestination." It actually has religious roots; in Christianity, it refers to whom God chooses to "save." In the more modern sense, predestination also includes an element of human moral will, but salvation is still up to God.[1] At first glance, this makes you think the plot of the movie is religious, but it's really not—it more talks about how some events seem to be fated or predestined to happen. I'll try to explain without getting too spoiler-y here.

Let's begin our time-traveling tale with a brief discussion of the inspiration for *Predestination*: "All You Zombies," a 1959 short story by Robert Heinlein. The screenwriters for *Predestination* say it's a

---

1   *Encyclopaedia Britannica*. (2019). Predestination. Retrieved from https://www.britannica .com/topic/predestination

story sixty years ahead of its time,[2] although other critics are not so kind; in a review for *Predestination*, AV Club called Heinlein's story "the most perversely preposterous addition to the genre" of time travel.[3] The movie follows the plot of "All You Zombies" very closely, in particular focusing on inserting paradoxes—things that cannot be explained by causality alone. In particular, the movie focuses on the "Bootstrap Paradox," which describes scenarios that are a bit like asking which came first—the chicken or the egg.[4] (By the way, in thinking back over other franchises we've covered already, you know that Bootstrap Paradoxes are super-common in fiction: *Terminator*, *Back to the Future*, and *Looper* are just a few examples in science fiction.) Cause and effect are broken apart in *Predestination* and it's a little hard to figure out which came first, and whether one can break the chain without breaking apart the universe.

Of course, revealing what those paradoxes are here would be a *major* spoiler, so I won't go too far into this—but just think about what happens in a typical life. We have expectations about the people around us, whom we love—as well as expectations about ourselves. "All You Zombies" explores what happens when we find out that we have been utterly betrayed. Is this tragedy something that actually forms an integral part of our identity, or can we rise above the circumstances? That's something teenagers ask themselves after every breakup, so it's interesting to see how similar scenarios play out in "All You Zombies," or *Predestination*, for that matter.

---

2 Russell, Stephen A. SBS Movies. (2016, Dec. 6). "Predestination: 5 Minutes with the Spierig Brothers and Sarah Snook (Interview)." https://www.sbs.com.au/movies/article/2014/08/01/predestination-5-minutes-spierig-brothers-and-sarah-snook-interview

3 Dowd, A.A. (2015, Aug. 1). "Predestination Gives A Famous Time-Travel Story the Chris Nolan Treatment." The AV Club. Retrieved from https://film.avclub.com/predestination-gives-a-famous-time-travel-story-the-chr-1798182343

4 Christoforou, Peter. "Predestination (2014) Explained." Astronomy Trek. Retrieved from http://www.astronomytrek.com/predestination-2014-explained/

Right, so how do people travel through time in Heinlein's story and in *Predestination*? Oh, that's simple to explain. A violin case. *What?* Okay, it's a little more than a violin case. It's something called a "USFF Coordinates Transformer Field Kit" that is in the shape of a violin case so as not to attract any attention from people seeing it being carried around. The movie script describes it as "a device that creates a temporal wake," in other words, a time machine. Fans of *Star Trek* will instantly know what I'm talking about here. The fan site Memory Alpha explains a temporal wake as "an energetic phenomenon that can be created as a byproduct of a temporal vortex, insulating anything caught inside from an alteration of the timeline."[5] So in other words, as long as you're close enough for the temporal wake to shield you, it acts as a barrier as time changes around you. A famous example of this happens in *Star Trek: First Contact*, when an alien species known as the Borg travels back in time to the year 2063 to assimilate (more practically speaking, annihilate) the human race. But the lucky USS Enterprise (with humans on board from the twenty-third century) goes in hot pursuit of the Borg ship just as it time warps. The Enterprise gets caught in the same temporal wake as the Borg, shielding it from the effects when the Borg changes history forever. And so a fun movie begins. I'm digressing, but hopefully this shows you how a temporal wake works.

How does time warping work in the first place? There is a physics concept called a "Lorentz transformation," but likely it doesn't have applicability here. Without getting too heavy into the mathematics, a Lorentz transformation is linked to the concept of general relativity. Equations from Dutch physicist Hendrik Lorentz show that, for example, time is relative when you travel

---

5   Memory Alpha. (n.d.) "Temporal Wake." Retrieved from https://memory-alpha.fandom
     .com/wiki/Temporal_wake

close to the speed of light.[6] One of the things that weirds out high school physics students is when they learn, according to Lorentz and Einstein's work, that as you travel close to the speed of light, time slows down for the person inside this theoretical spaceship, compared with their friends or family back on Earth. (Actually, the more proper way of explaining this is that time dilation occurs because moving through space alters the flow of time.[7]) So, you could leave Earth, hop in a spaceship, and tool around at the speed of light for a while, then arrive home to find out that you're about the same age—but your friends and family all aged and died in your absence. If you think about this, it *is* a form of time travel—but not in the way that *Predestination* describes it. When time dilation occurs, we experience a form of time travel into the future. But it can't be used to travel into the past. Perhaps the best way to explain *Predestination* is instead through ignoring the whole "transformation" idea and using that old-fashioned wormhole we've talked about so many times in this book. If physicist Kip Thorne can imagine creating a local wormhole that would easily fit inside of a room,[8] maybe that's the real physics happening in *Predestination* with its violin cases.

The Science of Identity

I've tried really hard not to get too heavy on the spoilers in these scientific reviews, but in the case of *Predestination* I need to give one big key thing away to discuss identity in more detail. If you don't want to learn this spoiler, then please skip to the next section. Otherwise, read ahead at your own risk!

---

6 *Encyclopedia Britannica.* (2019). "Lorentz Transformations." Retrieved from https://www.britannica.com/science/Lorentz-transformations

7 Zimmerman Jones, Andrew and Robbins, Daniel. "Slowing Time To A Standstill With Relativity." (n.d.) Dummies.com. Retrieved from https://www.dummies.com/education/science/physics/slowing-time-to-a-standstill-with-relativity/

8 Thorne, Kip S. (1995). "Black Holes & Time Warps: Einstein's Outrageous Legacy." New York: W.W. Norton.

As John explains to the bartender near the beginning of the film, John grew up female. "When I was a little girl . . ." is how John explains the situation, bringing the story back a few decades to John's childhood in an orphanage. Some reviewers have quickly leapt to conclusions about what gender John is now, or used to be.

John self-identifies as female at the beginning of John's life (the narration regularly repeats "girl"), and then a traumatic biological event occurs (which you'll learn more about in the movie). After that event, John's clear identity appears to shift. John's gender now is not clearly disclosed, although John describes the transition after that event this way: "I was no longer a woman, and I did not know how to be a man."

The definitions of sex and gender are different by culture and by situation. The best definition of gender probably comes from each individual's lived experience. An international body, the World Health Organization, boils down the distinction to this: "'Gender' describes those characteristics of women and men that are largely socially created, while 'sex' encompasses those that are biologically determined."[9] (Note this may not be a perfect definition as WHO's definition of "gender" is itself binary.) The unfortunate thing is not all genders are considered equally. At some times, this relates to our health, such as a 2016 study from Canadian researchers showing that blood flow blockage to the heart may be related to gender characteristics.[10] But in most cases, genders face societal obstacles that can lead to poorer outcomes in employment, health, daily safety, and relationships, among many other factors. This is especially true of genders that are not considered in the majority.

9　World Health Organization. (2019). "Genomic resource centre." Retrieved from https://www.who.int/genomics/gender/en/

10　Conger, Krista. (2017). "Of mice, men and women." Stanford Medicine. Retrieved from https://stanmed.stanford.edu/2017spring/how-sex-and-gender-which-are-not-the-same-thing-influence-our-health.html

If there is a greater conclusion to be drawn from scientific discussions of gender, it is that part of science's role is to act in a neutral, fact-based capacity in areas of high political and religious discussion. To this day, some politicians and other "thought leaders" continue to advocate for a heterosexual society because that's the way it was always done in Eurocentric cultures they cite. However, science shows us that gender is not that simple—our world can't work that way. This unfortunately is not a debate that will change in a decade, or even in a generation. When I was a young journalist being trained in university, one of the things our instructors drilled in to us was our responsibility to speak about the underrepresented, because they are not offered the privilege of a platform in the same way that a majority is. The greater lesson for scientists reading the news? There is an inherent cultural bias toward representing the majority of the population, because it's easy to do. But that's not the right thing to do. It's my sincere hope that more science fiction platforms will include gender in their discussions, because fiction helps us better make sense of our world. And by making greater worlds in fantasy, hopefully our real-life world will improve as well.

The Legacy of *Predestination*

Predestination is one of those movies that explores the concept of cycle, which is by no means a new concept. As Futurism suggests, the ultimate question of cycles is the fate of the universe.[11] Some scientists suggest that there is enough dark matter in the universe to make expansion stop dead, and then everything collapses back in on itself in a morbid reversal of the Big Bang. But the majority

---

11 Monetti, Rich. (n.d.) "Predestination will Throw and Twist You Through Time and Space." Futurism. Retrieved from https://futurism.media/predestination-will -throw-and-twist-you-through-time-and-space

believe that our universe will expand forever, getting colder and larger until even fundamental molecules are torn apart.

Cycles are part of the rhythm of life on Earth. The sun, moon, and planets rise and set in our sky on predictable schedules. Many of us have regular seasons (at least, until climate change messes that up for good). Our own lives go through cycles: when we are young, we learn from older people, and when we are old we teach younger people. Then there are those religions or worldviews or mythologies that say we may be in a cycle of birth, death, and reincarnation. In *Predestination*, it's not clear if this is a natural cycle or a cycle of corruption. As you see some various shadowy characters identified, your sense of time gets all distorted. I found myself wondering, could what is happening on screen actually physically happen? Or is it just a story plot to explore the evolution of an individual's identity over time? The complexity of the story makes you want to view it more than once, because it's hard to pick apart individual motivations. And that's baked right into the screenplay from the beginning. "We spent a couple of years working on the script and trying to ensure every word, every line, every moment has value and has a purpose and has relevance," said Michael Spierig, who cowrote the screenplay with his brother, Peter. "And it may seem, perhaps, on first viewing that some of it may not, but it all does. And we worked really hard on ensuring that that was the case."[12]

In thinking about this film, I think that cycles can also be true when thinking of how to identify ourselves. I'm approaching my forties as I write this book. In university, I had a certain perception of myself, a person who enjoyed writing about science, but who struggled with my self-identity as feeling awkward and nerdy. For example, I knew far more about the Apollo moon landings or *Star*

---

12 Creative Planet Network. "The Spierig Brothers Talk Their Repeat Viewing Time Travel Film 'Predestination'." Retrieved from https://www.creativeplanetnetwork.com/news-features /spierig-brothers-talk-their-repeat-viewing-time-travel-film-predestination-608079

Trek than a lot of folks in my journalism program. I was proud of that knowledge, but I wasn't sure if it was the "right" knowledge to have in university. Some fifteen years later, I look back at my younger self and I can see the seeds of who I am today. I now *exploit* the very nerdy and awkward traits that used to drive me crazy—heck, I wouldn't be writing this book without them. Nor would I be married to the person I married (who is awesomely nerdy in his own way—I mean, we first bonded over a *radio telescope*, of all things). And I see my job now as encouraging those traits in younger people.

So, that's a cycle, right? My ultimate dream is trying to show younger people that they should follow in careers similar to mine to share their love of science and to encourage the next generation to learn about science, which in turn would encourage a yet-unborn generation to learn about science . . . and our world hopefully keeps progressing and getting more nerdy with each generation. (As long as climate change doesn't kill us first.)

X-MEN: DAYS OF FUTURE PAST (2014)

It doesn't matter how many times Hugh Jackman puts on Wolverine's claws—he still remains a compelling actor to watch. In this movie, his character Wolverine has his consciousness sent back to 1973 to meet the younger versions of Magneto (Michael Fassbender) and Professor X (James McAvoy), and some of the other X-Men battle a class of huge mutant-killing robots called the Sentinels.

The Science of Time Travel

To understand how Wolverine went back in time, we need to briefly explain what the X-Men are; note that despite its name, not all the X-Men are men. The X-Men was originally a Marvel Comics comic book series that started publication in 1963, by Stan Lee (writer) and Jack Kirby (cowriter and artist). This group is largely based on mutants with superhuman abilities, thanks to an X-Gene that seems to promote different abilities in different people. Naturally, these abilities give the X-Men the ability to supersede human power and strength, but for the most part, they use it for good—to keep humans and mutants living happily together even though there is a lot of racism against mutants.

Of course, just like people, not all mutants are on the same side of the debate. The X-Men's leader is Charles Xavier (a.k.a. Professor X) who is a mind-reader and controller. Against them is the Brotherhood of Mutants leader Magneto, who has a cool ability to influence magnetic fields. Magneto believes that mutants should fight against humans, but sometimes he'll team up with the

X-Men if they have a mutual cause to work toward. The X-Men is still running as a comic book series and there have been a number of movies, too, including the current series that started way back in 2000.

So, why are we going back in time this time? In the year 2023, the Sentinels trap the X-Men in a Chinese temple and try to break in to kill the small band of mutants. While some of the X-Men fend the Sentinels off outside, the group inside tries to think their way out of the situation. Xavier gives a little history lesson on how the Sentinels were formed—essentially, these robots really got a technological advance after a mutant called Mystique killed their designer (Bolivar Trask). Unhappy government officials captured Mystique and extracted her DNA to make the Sentinels even more powerful than Trask's design, leading to our little dilemma now in 2023.

Naturally, the easiest thing—instead of fighting them off, Xavier says, why not go back in time? Unfortunately, even a mutant will die from the stress and shock, unless you're Wolverine and able to regenerate on the spot. And being mutants, they don't need no stinking time machine to accomplish time travel. Instead, Wolverine lies down and allows his mind to fall under the influence of Kitty Pride, who sends his consciousness back to 1973. *What?* We only use our *minds* to go back through time in this movie?

Well, first of all, you and I do it every day. That's right, every time we think about a memory from the past and use it to try to predict the future, that's what psychologists called mental time travel (or if you prefer the more scientific sounding term, chronosthesia). The first mention of this kind of time travel was published in 1997, by Thomas Suddendorf and Michael Corballis,[1] although it was Endel Tulving[2] who created the term chronosthesia.

---

1   Suddendorf, T. and Corballis, M.C. (May 1997). "Mental time travel and the evolution of the human mind". *Genetic, Social, and General Psychology Monographs* 123(2): 133–67
2   Tulving, E (2002). "Chronosthesia: Conscious Awareness of Subjective Time". *Principles of Frontal Lobe Function*. 311–325.

Psychologists have a few debates about the nature of chronosthesia. For example, is it a uniquely human thing, or do animals possess it too? I guess that depends on whether you think the tools that some large apes use to pick up food is a result of foresight, or just of pure instinct. Other questions include when children pick up this skill, and how our ability to "see" the future is related to old philosophical debates such as free will. If you know you panic every time you see a spider, for example, how much can you control that the next time it happens?

But wait, chronosthesia is not time travel in the physical sense. It's just memories. Heck, as much as we can try to imagine ourselves in 1973 in our minds, we're not actually living it—able to alter its future. Is there a way to actually bring ourselves back a few decades, as Wolverine agreed to do? To be honest, short of using the same old wormhole trope or creating some kind of a time machine that would move your brain . . . it's hard to imagine any way to literally go back to 1973 using your mind. (Unless you're a mutant, of course—who knows how Wolverine's brain works.)

The Science of Flow

What I want to show you today, though, is that you can use your mind to alter time in ways you did not expect—and no, I don't mean by using hallucinatory drugs. I mean by using the concept of flow. This was developed by Hungarian-American psychologist Mihaly Csikszentmihalyi. He spent years outfitting people with beepers that went off at irregular intervals, then had them answer questions about what they were doing and how they were feeling at the time.

As he explains in the beautiful book *Flow: The Psychology of Optimal Experience* (1990)—I've read it many times and highly recommend it—flow requires a curious combination of elements. Say you're playing tennis. You need to be fully focused on your task, so please don't use your smartphone and play tennis at the same time. You need to be doing something that is very slightly challenging, but not *too* challenging (so don't have Serena Williams as your opponent). You need to set a goal and have clear steps along the way, perhaps aiming to hit the tennis ball to a certain area on the court. If you focus intensely, eventually you will feel flow—a state where hours feel like minutes and that when you look up from your task, you can't believe how much time has passed. Csikszentmihalyi argues that flow is a way to achieve happiness—perhaps not happiness in the immediate sense, like marathoning that series on Netflix, but happiness in the sense that you are growing and learning and contributing something for yourself or the world at large. The famous Internet comic, The Oatmeal, describes author Matthew Inman's struggles with depression and how he uses flow to accomplish things.[3] He works for twelve hours at a stretch, "until I smell weird." He runs fifty miles consecutively, "over mountains until my toenails fall off." He admits that he doesn't feel happy during these times—in fact, he's "often suffering." But he continues, "I do them because I find them meaningful. I find them compelling." What's your next compelling action to alter time?

Another great feature of flow is, in these days of consumerist tendencies, achieving flow doesn't require a lot of money. What it *does* need is time and energy and a space to concentrate, which unfortunately are highly associated with people of privilege. But if you can, get to the library and read that book about computer programming that interests you. Join a community class on watercolors to figure out how to best capture the light streaming on the flowers by your bedside window. And when you do watch TV, because most of us do, mix in your *Game of Thrones* marathons with documentaries that challenge you, or YouTube channels with educational

---

3   Inman, Matthew. (n.d.) How to be perfectly unhappy. The Oatmeal. Retrieved from https://theoatmeal.com/comics/unhappy

content. (My personal favorite? *Crash Course*—a project of Hank Green and John Green to provide basic education in all subjects at the high school level. It's cute and you find yourself learning unconsciously, guided by their humor and flair. Do check it out.)

The Legacy of X-Men: Days of Future Past

Will this movie leave a lasting impact on viewers? *Vulture* argues that the time has long passed for that. We are so inundated with post-apocalyptic movies and superheroes these days that X-Men tends to blend into the noise. The 1981 comic inspiration for this movie, however—that's something different. It was an era where alternate realities weren't easily accepted. Today we're used to doozies such as *Arrival* (I still can't figure out the mechanics of that Chinese phone call) and *Lost* (are they all dead, or what?). But this was an era before *Back to the Future*, before even *Terminator*. And remember, the parents of these comic book readers grew up in a time when the TV screen needed to ripple and "handhold" viewers to signify a flashback.[4] So readers felt disjointed and disturbed when, without any prelude to how they arrived there, in *Uncanny X-Men No. 141*, the X-Men suddenly found themselves in a bombed-out New York City fighting Sentinels. Let's just say (based on *Vulture*'s account) that things went terribly.[5] And creators remembered. One famous cover with mugshots of the X-Men, blazoned with the words "slain" or "apprehended" on each one, influenced a who's-who of comic franchises in decades afterward: *Doctor Who*, *Star Wars*, even the Canadian *Alpha Flight*.

4　Johnson, Steven. (2018, Nov. 15). "The Human Brain is a Time Traveler." *The New York Times Magazine*. Retrieved from https://www.nytimes.com/interactive/2018/11/15/magazine/tech-design-ai-prediction.html

5　Rieseman, Abraham. (2014, May 22). X-Men's 'Days of Future Past' Is One of the Most Influential Comics Stories Ever. Here's How It Happened. *Vulture*. Retrieved from https://www.vulture.com/2014/05/x-men-days-of-future-past-origins-legacy.html

The original concept came to be when writer Chris Claremont and artist John Byrne realized that their four-year partnership was about to come to a close; after building up X-Men since 1977, Byrne was reassigned to *Fantastic Four*. The duo decided to end their collaboration with a bang. They had recently killed off one X-Men character and got hate mail and huge sales at the same time, prompting Byrne to speculate a great idea would be just to kill. them. all. One of the covers of the series showed Wolverine being burned to death, with the tag line: "This issue: Everybody dies!" Of course, the writers found a way out of this situation eventually by time traveling the heck of the X-Men universe—a move that may be confusing to new fans as multiple versions of X-Men stroll around the comic books simultaneously.[6]

Looking beyond the franchise, it's clear that both the X-Men movies and the comic books are about prejudice, especially among minority groups or those with different abilities. While I like to think our sensibilities have evolved in past decades as to how to treat others, at times the news is worrisome. Immigrant children in cages at the US-Mexican border, northern indigenous people facing a lack of drinking water, and Ontario kids with autism affected by government cutbacks to treatment, were all issues of great discussion in Canada (my home country) in early 2019, as I wrote this book. That's not even to mention the historical issues of colonialism, slavery, oppression against women and people of color and those of non-heterosexual gender, and a host of other issues that may have happened generations ago. Yes, these still affect people today. A single example: If you're a twentysomething African American, it's probable that your grandparents were unable to buy a house in a good neighborhood just after the Second World War. Those who did have the advantage three generations ago built

---

6   *Ibid.*

up generational wealth allowing their kids and grandkids to go to better schools and thus, get better jobs, the argument often goes.

What can one person do against a culture that to this day, privileges the rich and the powerful? Actually, there's a lot you can do by your own self. Help your neighbors, give to community organizations, teach your children and friends how to be respectful, and above all—to listen. To accept that somebody's lived experience has value and should be honored. To set an example with your own hands and heart, and to vote carefully and responsibly whenever you are called upon to do so.

ACKNOWLEDGMENTS

My biggest debt goes to the creators of the time-travel stories, and all the fans and observers over the years who have analyzed their work. I learned a lot from reading everything, and I hope to do justice to some of the discussions in this book.

Skyhorse Publishing was an enthusiastic backer from the start. In particular, thanks to Nicole Mele for helping me where I fudged things, and to Jason Schneider for inviting me to take on the book. The team behind them who helped with the publishing and promotion is a marvel to work with.

My virtual assistant, Christina Goodvin, did a lot of the background research. She's whip-smart and I wish I could employ her full-time—she is a joy to work with.

As a freelancer, I kept working for money while writing this book. I truly have the best clients who understood when I rearranged a few meetings, or left emails aside, while delving into some science fiction fun. My part-time employer, Algonquin College, is very supportive with book-writing professors; their schedule flexibility is something I am very grateful for. Thanks to my department and to my students for understanding my occasional brief leaves of absence for my work.

Thank you also to the numerous family members and friends who put up with a very busy me in 2019 (first with a PhD, and second with multiple book-writing projects that included this one). First among those people— in their support and in my heart— is my husband J, who literally sat beside me many a time as I struggled to put my thoughts into words. I couldn't find a better partner in life or work, even if I could time travel through the ages. So thanks for believing in me.

A better world starts with our own family, our own lives. Do what you can for good. We need you.

INDEX